Fixed Point Theory for Decomposable Sets

Topological Fixed Point Theory and Its Applications

VOLUME 2

Fixed Point Theory for Decomposable Sets

by

Andrzej Fryszkowski

Faculty of Mathematics and Information Science,
Warsaw University of Technology, Warsaw, Poland

KLUWER ACADEMIC PUBLISHERS
DORDRECHT / BOSTON / LONDON

A C.I.P. Catalogue record for this book is available from the Library of Congress.

ISBN 978-90-481-6672-5 (PB)
ISBN 978-1-4020-2499-3 (e-book)

Published by Kluwer Academic Publishers,
P.O. Box 17, 3300 AA Dordrecht, The Netherlands.

Sold and distributed in North, Central and South America
by Kluwer Academic Publishers,
101 Philip Drive, Norwell, MA 02061, U.S.A.

In all other countries, sold and distributed
by Kluwer Academic Publishers,
P.O. Box 322, 3300 AH Dordrecht, The Netherlands.

Printed on acid-free paper

Printed in the Netherlands

Many thanks to professor Lech GÓRNIEWICZ who was encouraging me to prepare this monograph.

Contents

Introduction

Decomposable sets were introduced by T. R. Rockafellar [205] in 1968 and since then they have became one of basic objects in nonlinear analysis, especially in multivalued mappings theory. A subset K of measurable functions is called decomposable iff

$$(Q) \qquad \chi_A u + (1 - \chi_A) v \in K$$

for all $u, v \in K$ and measurable A. The property (Q) looks like the convexity condition and it turned out to be a substitute of convexity in many problems connected with optimal control, differential inclusions and closed areas. While the most of existence problems for convex case can be deduced from the Schauder Fixed Point Theorem, Banach Fixed Point Theorem, Kakutani Fixed Point Theorem and others, so the lack of convexity assumptions certainly required a developing of an adequate fixed point techniques and have engaged many mathematicians all over the world (some names will be mention in the literature, but the list is far from being complete). It turned out that many previously obtained results have analogues in "decomposable analysis". Our book is an attempt of showing the present stage of this topic. We are especially directed towards the theory of fixed points for decomposable sets since they lead to the most of important applications in the nonlinear analysis.

This survey is addressed to mathematicians and students working out in topology, geometry optimal control, differential inclusions and functional analysis. It will require a good background from the functional analysis and topology. The material contained is split into three parts. The first one is devoted to a background from functional analysis, the second to the theory of multifunctions and the third to the decomposability property. Most of the presented facts connected with fixed points and their applications is closer or further consequence of the Michael Selection Theorem ([157]) and it's "decomposable" version - theorem due to Bressan-Colombo and Fryszkowski ([35] and [72]). Both selection results are clipped by the Lapunov Convexity Theorem

([**140**]) (on the range of a vector measure) and it's generalizations, including the Aumann integrals. To give an appriopriate language and proper tools we start, in chapter I, with a brief recollection of the convex sets theory, as well in arbitrary Banach spaces as in euclidean ones. We invoke the Carathéodory Theorem and give it's generalization due to Olech [**171**].

In the second chapter we deal with vector measures introducing the notion of essential extremum of an arbitrary family of nonnegative measures and describing the segments and ε-segments for nonatomic measures. The segments allow us to construct in arbitrary measure space (T, Σ, μ), via the Liapunov-like results, a similar structure as in the interval $[0, 1]$. This, in turn, leads to a continuous partition of T on the parts proportional (or almost proportional) to a prescribed partion of the unity. The continuous partition techniques we use in the chapter IX deducing a decomposable version of continuous selection theorem for lower semicontinuous multifunctions. The convex case - Michael Selection Theorem - is discussed in the chapter IV.

We have also to pay the reader's attention that continuous selections are somewhat tied with lower semicontinuity of multivalued mappings assuming convex or decomposable values. In other regularity types continuous selections may fails to exists. But sometimes they can be replaced by measurable selections. Questions connected with this kind of problems we have discussed in the chapter V. In particular there is presented the Kuratowski & Ryll-Nardzewski Selection Theorem ([**139**]) and it's complement - the Castaing representation of measurable multifunctions ([**42**]). In the chapter VI and XII we combine previously obtained results to the Carathéodory type mappings, i.e. mappings of two variables, measurable in the first one and semicontinuous (lower or upper) in the second. A reason of the considering such objects comes from the differential inclusions, where the right-hand side are, in general, multivalued mappings of two or more variables. The differential inclusions we discuss in the chapter XIV. From the point of view of the fixed point theory the most grateful tasks are related with Lipschitz differential inclusions and multivalued contractions. We explain in details the continuous version of the Filippov Lemma and the Filippov-Ważewski Relaxation Theorem for differential inclusions depending on a parameter. Provided considerations lead to the existence of continuous (with respect to the parameter) selections not only for the solutions sets to differential inclusions but also for a multifunction describing (for a given parameter) all possible derivatives. Complementarily, each set of derivatives is, in the case of Lipschitz differential inclusions, an absolute retract. Saying more precisely, the set of fixed points of a multivalued contraction is a retract of the space of integrable functions. This is a statement of the theorems due to Ricceri [203] - in convex-valued case and Bressan, Cellina and Fryszkowski [36] - the decomposable-valued ones. We are obliged to emphasize that obtaining both results required a laborious developing of the theory of multivalued mappings and the fixed point theory.

The topic of the last chapter XV is related with decomposable mappings and integral functional given by such kernels. There are given necessary an sufficient conditions for the lower semicontinuity of such functional in certain spaces of integrable functions. In this part of material we do not apply fixed point technics. But it requires a strong understanding of the nature of decomposable sets. A consideration of this class of mappings can lead to an unified theory of the continuity of the integral functionals.

During the preparation of this work I have been helped significantly by a number of friends and colleagues. Special gratitude we owe to dr Grzegorz Bartuzel who has carefully read the manuscript contributing his advice and criticism. I would like also to thank Mrs Marlies Vlot from the Kluwer Academic Publishers for her patience during the whole preparation period.

Warsaw, 2004

Andrzej Fryszkowski

Part 1

FUNCTIONAL ANALYSIS
BACKGROUND

CHAPTER 1

Preliminaries

The main objects examined in this monograph are multivalued mappings (multifuntions) and their fixed points. Let T and X be given sets. By a multivalued mapping, or, in other words, a multifunction we mean a mapping $P : T \longrightarrow N(X)$, where $N(X)$ stands for a family of all subsets of X. Of course any single-valued function $p : T \longrightarrow X$ can be identified with the multivalued mapping $P(t) = \{p(t)\}$. Multifunctions are denoted by capital letters P, R, F, G, L, ... These objects appear in a natural way in many different situations. Below we give some examples of them:

EXAMPLE 1. $P(x) = \{y : w(x, y) = 0\}$, where $w(x, y)$ is a polynomial of two variables;

EXAMPLE 2. $F(t) = f(t, Y)$, where $f : T \times Y \longrightarrow X$ be a given function;

EXAMPLE 3. $G(t) = \{x : g(t, x) \leq 0\}$ for given $g : T \times X \longrightarrow R$;

EXAMPLE 4. $Log(z) = \{w : \exp w = z\}$ defined on $\mathbf{C} \setminus \{0\}$.

By the Axiom of Choise for every multivalued mapping $P : T \longrightarrow N(X)$ one can pick up a function $p : T \longrightarrow X$, such that $p(t) \in P(t)$ for all $t \in T$. Such a function is usually called a selection of P. If additionally p is measurable, continuous or Lipschitz we say that p is measurable, continuous or Lipschitz selection.

A point $x_0 \in X$ is said to be a fixed point of a multivalued mapping $P : X \longrightarrow N(X)$ if

$$x_0 \in P(x_0).$$

This notion is a generalization of fixed points for single-valued functions $p : X \longrightarrow X$. Among many problems concerning fixed points we shall discuss the existence and their properties. In the examination of multifunctions and their fixed points a crucial role play properly chosen selections. Namely, any fixed point of a selection p is a fixed point of P. There are many existence results of fixed points concerning multifunctions and most of them are analogues of single-valued case. Some of them require convexity assumptions on the values of multifunctions,

3

cf. Kakutani [**126**], Ky Fan (cf. [**8**], [**9**], [**116**]), the Banach contraction principle ([**62**]), Ricceri [**203**], some do not - Bressan-Cellina-
Fryszkowski [**36**]. Our work does not pretend to give all of them and we
are mostly interested in those which have direct roots with functional
analysis. We have to say that topological and algebraic approach is
not even touched, however it is an important branch of the fixed point
theory.

Described in the present book theory of fixed points has many applications in widely understood analysis, especially in differential inclusions. By a differential inclusion in a functional space X we mean a
relation

$$(1.1) \qquad\qquad Dx \in F(x),$$

where D is a differential operator. A straightforward application of
fixed point theory to problem (1.1) comes from an easily varified fact:

Assume that the operator D admits a right inverse R. Then $x = Rs$
is a solution of (1.1) iff s is a fixed point of the mapping $K(s) = F(Rs)$,
i.e. $s \in K(s)$.

As an example of differential inclusions the reader may think of the
relation $x'(t) \in F(t, x(t))$, where $t \in I = [0, 1]$, $F : I \times R^l \longrightarrow N(R^l)$
and $x : I \longrightarrow R^l$ is an absolutely continuous function. In this case
$D = \frac{d}{dt}$ and denoting $s = x'$ we have

$$x(t) = R(s)(t) = \zeta + \int\limits_0^t s(\tau)\, d\tau.$$

In this case
(1.2)
$$K(s) = \{u : I \longrightarrow R^n |\ u \text{ is integrable and } u(t) \in F(t, R(s)(t))\ a.e.\}.$$

If the values of F are convex then K is convex-valued as well and to
existence results for such (1.1) we just need fixed points for a convex-
valued multifunction. But the convexity assumption for the sets $F(t, x)$
is in many situations artificial, especially in nonlinear control theory or
in differential inclusions. A glance on (1.2) shows that it posseses the
following property:

for every $u, v \in K(s)$ and any measurable $A \subset I$ the function

$$(Q) \qquad\qquad \chi_A u + (1 - \chi_A) v \in K(s).$$

This condition is called decomposability and occurs to be a proper tool
in examination of arbitrary multifunctions.

1. Basic notations and symbols

Let X be a topological space. For a given $A \subset X$ by clA and $IntA$ we mean, respectively, its closure and interior. In X we shall distinguish the following families:

$$N(X) \quad - \quad nonempty\ subsets;$$
$$cl(X) \quad - \quad closed\ nonempty\ subsets;$$
$$c(X) \quad - \quad compact\ nonempty\ subsets.$$

Let (X, d) be a metrizable space with metric d. An open ball centered at $x \in X$ and radius r we shall denote by $B(x, r)$, while the symbol $\overline{B}(x, r)$ stands for the closed one. In (X, d) we also consider the families:

$$b(X) \quad - \quad bounded\ nonempty\ subsets;$$
$$cl(X) \quad - \quad closed\ nonempty\ subsets;$$
$$bcl(X) \quad - \quad bounded\ closed\ nonempty\ subsets.$$

The distance between a set $A \subset X$ and a point $x \in X$ we shall denote by

$$(1.3) \qquad d(x, A) = \inf\{d(x, a) : a \in A\}$$

Let us recall that

$$d(x, A) = d(x, clA)$$

and the mapping $x \longrightarrow d(x, A)$ is continuous. Moreover, if $A \in N(X)$ then it is Lipschitz with constant 1.

Let $A, B \in b(X)$ be given. By the Hausdorff distance between A and B it is usually meant

$$d_H(A, B) = \max\{d_0(A, B), d_0(B, A)\},$$

where $d_0(A, B) = \sup_{a \in A} d(a, B)$.

PROPOSITION 1. $(b(X), d_H)$ is a metric space and $(c(X), d_H)$ is complete.

Having a metric space (X, d) we can also consider the space $\mathcal{C}(S, X)$ of continuous functions f defined on a Hausdorff topological space S with values in X. This space is metrizable by

$$(1.4) \qquad \mathbf{d}(f, g) = \sup_{s \in S} \arctan d(f(s), g(s))$$

and within $\mathbf{d}$ it is complete whenever such is X.

Now let X be a Banach space with norm $|\cdot|$ and conjugate X^*. Unit balls in X and X^* we shall denote, respectively, by B and B^*.

For $x^* \in X^*$ the dual pair is $\langle x^*, x \rangle = x^*(x)$. Recall that X^* is a Banach space with the norm

$$|x^*| = \sup \{ \langle x^*, x \rangle : x \in B \}.$$

By the Banach-Alaoglu Theorem the unit ball B^* endowded with *weak*− topology* is a compact set. Moreover, if X is separable then B^* is metrizable by metric

$$d^*(x^*, y^*) = \sup_n \{|\langle x^* - y^*, x_n \rangle|\},$$

where $\{x_n\}_{n=1}^{\infty}$ is a dense subset in B.

In this case $\mathcal{C}(S, X)$ is a locally convex metrizable space. Additionally the set $C(S, X) \subset \mathcal{C}(S, X)$ of continuous an bounded functions f is a Banach space with norm

(1.5) $$\|f\|_{\infty} = \sup_{s \in S} |f(s)|$$

and the norm topology is equivalent to the metric one. We also have to recall that for Hausdorff compact topological space S we do not need to assume the boundedness of any continuous function $f : S \longrightarrow X$, since this property we have "for free". As an example we have the space $C(I, X)$, where $I = [0, 1]$.

2. Convex sets

Let X be a vector space. A set $A \in N(X)$ is said to be convex iff for every $a, b \in A$ and $\lambda \in [0, 1]$ we have

$$\lambda a + (1 - \lambda) b \in A.$$

For given $A \subset X$ the symbols coA and $clcoA$ stand for, respectively, convex hull and closed convex hull of A. For any $a, b \in X$ the set $co\{a, b\}$ we denote by $[a, b]$ and call it the segment joining a and b. We shall also consider the following families:

$$
\begin{array}{lll}
co(X) & - & \textit{convex nonempty subsets;} \\
clco(X) & - & \textit{convex nonempty subsets;} \\
cb(X) & - & \textit{bounded closed convex nonempty subsets;} \\
cc(X) & - & \textit{compact convex nonempty subsets.}
\end{array}
$$

A special class among convex sets represent convex cones. A set $\times \in N(X)$ is called a cone iff for every $c \in \times$ and any $\lambda \geq 0$ the point $\lambda c \in \times$. If additionally $\times$ is convex then we say that it is a convex cone. We shall only consider cones $\mathcal{C}$ possessing the following properties:

(1.6) $$\times \cup (-\times) = R^l, \quad \times \cap (-\times) = \{0\}.$$

Such cone $\times$ induces the order by saying

$$a \leq_{\times} b \quad iff \quad b - a \in \times.$$

The reader may have already noticed that for given cone $\times$ any two elements $a, b \in X$ are, by (1.6), comparable. So it is justified to introduce lexicographical extrema of a set point b is called a lexicographical maximum of $W \in N(X)$ if

   i. for each $w \in W$ we have $w \leq_{\times} b$

     and

  ii. if for some c and all $w \in W$ there is $w \leq_{\times} c$ then $b \leq_{\times} c$.

Such point b, if exists, is unique and we denote it by $\max(W, \times)$. Similarly we define the lexicographical minimum $\min(W, \times)$ of W. Obviously

$$\min(W, \times) = -\max(W, -\times).$$

Generally speaking in examination of convex sets or convex hulls an useful language provide support functions. For a set $A \in N(X)$ the support function $c_A : X^* \longrightarrow R \cup \{\infty\}$ is defined by

$$c_A(x^*) = \sup\{\langle x^*, x \rangle : x \in A\}.$$

PROPOSITION 2. *Given $A \in N(X)$. Then:*

  1. c_A *is a convex, positively homogeneous function and*

$$c_A = c_{coA} = c_{clA} = c_{clcoA}.$$

  2. for bounded A the mapping $x^* \longrightarrow c_A(x^*)$ is Lipschitz with constant

$$M = \sup\{|x| : x \in A\} < \infty.$$

  3. for $A, B \in b(X)$ we have

(1.7) $$\sup_{|x^*| \leq 1} |c_A(x^*) - c_B(x^*)| \leq d_H(A, B).$$

PROOF. The first statement is obvious. To see 2. fix $x^*, y^* \in X^*$ and $\varepsilon > 0$ and take any $a, b \in A$ such that

$$c_A(x^*) < \varepsilon + \langle x^*, a \rangle \quad and \quad c_A(y^*) < \varepsilon + \langle y^*, b \rangle.$$

Then

$$c_A(x^*) - c_A(y^*) \leq \varepsilon + \langle x^*, a \rangle - \langle y^*, a \rangle \leq \varepsilon + M\|x^* - y^*\|.$$

Similarly

$$c_A(y^*) - c_A(x^*) \leq \varepsilon + \langle y^*, b \rangle - \langle x^*, b \rangle \leq \varepsilon + M\|x^* - y^*\|$$

and therefore

$$|c_A(x^*) - c_A(y^*)| \leq \varepsilon + M\|x^* - y^*\|.$$

Now passing to the limit with $\varepsilon \longrightarrow 0$ yields the required property.

For the third statement fix x^* with $|x^*| \leq 1$ and $\varepsilon > 0$. Take any $a \in A$ and pick such point $b \in B$ that

$$|a - b| < d(a, B) + \varepsilon$$

Therefore

$$\langle x^*, a \rangle - c_B(x^*) \leq \langle x^*, a \rangle - \langle x^*, b \rangle \leq |a - b| < d(a, B) + \varepsilon$$

and thus

$$c_A(x^*) - c_B(x^*) \leq \sup_{a \in A} d(a, B) + \varepsilon \leq d_H(A, B) + \varepsilon.$$

Since the role of A and B is symmetric then we conclude that

$$|c_A(x^*) - c_B(x^*)| \leq d_H(A, B) + \varepsilon$$

and allowing $\varepsilon \longrightarrow 0$ we get (1.7).   □

For convex sets an important role plays the following:

PROPOSITION 3. *The closed convex hull of $A \in N(X)$ can be represented as follows:*

$$clcoA = \bigcap_{x^* \in X^*} \{x : \langle x^*, x \rangle \leq c_A(x^*)\} = \bigcap_{x^* \in B^*} \{x : \langle x^*, x \rangle \leq c_A(x^*)\}.$$

If, additionally, X^* is separable and A is bounded then for any dense subset $\{x_n^*\}_{n=1}^{\infty} \subset B^*$ we can write

$$(1.8) \qquad A = \bigcap_{n=1}^{\infty} \{x : \langle x_n^*, x \rangle \leq c_A(x_n^*)\}.$$

The above proposition can be extended for closed convex hull in so called $\mathcal{Z}-$ topology in X.

DEFINITION 1. *A linear subspace $\mathcal{Z} \subset X^*$ is called total if condition $\langle x^*, x \rangle = 0$ for all $x^* \in \mathcal{Z}$ implies $x = 0$.*

Any total space $\mathcal{Z}$ determines $\mathcal{Z}-$topology in X by introducing a basis of neighbourhoods

$$V(x_0, z_1, ...z_n, \varepsilon) = \{x : |\langle z_i, x - x_0 \rangle| < \varepsilon, \ i = 1, ..., n\}$$

of any $x_0 \in X$, where $z_1, ...z_n \in \mathcal{Z}$. The convergence of a net $\{x_\alpha\}$ in $\mathcal{Z}-$topology to x_0 we denote by $x_\alpha \xrightarrow{\mathcal{Z}} x_0$. It means that for each $z \in \mathcal{Z}$ we have $\langle z, x_\alpha \rangle \longrightarrow \langle z, x_0 \rangle$.

For $A \in N(X)$ denote by $clco_{\mathcal{Z}}A$ the smallest $\mathcal{Z}-$closed and convex set containing A.

PROPOSITION 4. *For any $A \in N(X)$ the formula*

$$clco_{\mathcal{Z}} A = \bigcap_{z \in \mathcal{Z}} \{x : \langle z, x \rangle \leq c_A(z)\}$$

holds.

PROOF. Denote the right-hand side in the above formula by A_0. Evidently A_0 is convex. It is also closed. To see this take arbitrary $x_\alpha \xrightarrow{\mathcal{Z}} x_0$ with $\{x_\alpha\} \subset A_0$. Since for every $z \in \mathcal{Z}$ we have $\langle z, x_\alpha \rangle \leq c_A(z)$ then $\langle z, x_0 \rangle \leq c_A(z)$, what implies that $x_0 \in A_0$. Therefore $clco_{\mathcal{Z}} A \subset A_0$.

To see the opposite inclusion take any $x_0 \in A_0$ and let

$$V = V(x_0, z_1, ... z_n, \varepsilon) = \{x : |\langle z_i, x - x_0 \rangle| < \varepsilon, \ i = 1, ..., n\}$$

be an arbitrary neighbourhood of x_0. We need to show that $V \cap A \neq \emptyset$. For this purpose consider the set

$$W = \{w = (w_1, ..., w_n) \in R^n : w_i = \langle z_i, x \rangle, \ x \in A\}.$$

By the Proposition 3

$$clcoW = \bigcap_{d \in R^n} \{w : \langle w, d \rangle \leq c_W(d)\}.$$

But

$$c_W(d) = \sup\left(\left\langle \sum_{i=1}^{n} d_i z_i, x \right\rangle : x \in A\right) = c_A\left(\sum_{i-1}^{n} d_i z_i\right).$$

Hence $w_0 = (w_{10}, ..., w_{n0})$ with $w_{i0} = \langle z_i, x_0 \rangle$ is a member of $clcoW$. Indeed, for every $d = (d_1, ..., d_n)$ we have

$$\langle w_0, d \rangle = \sum_{i=1}^{n} w_{i0} d_i = \left\langle \sum_{i=1}^{n} d_i z_i, x_0 \right\rangle \leq c_A\left(\sum_{i=1}^{n} d_i z_i\right) = c_W(d).$$

Thus for any $\varepsilon > 0$ there exists $x \in A$ such that

$$|\langle z_i, x - x_0 \rangle| < \varepsilon, \ i = 1, ..., n$$

and this is nothing else that $x \in V \cap A$. This completes the proof. $\square$

Each compact and convex subset in X (and in any locally convex topological space) can be characterized by extreme points. Recall that e is said to be an extreme point in $A \subset X$ iff the representation

$$e = \lambda a + (1 - \lambda) b$$

for some $a, b \in A$ and $\lambda \in I$ can hold only for $a = b = e$. The set of extreme points of $A \in N(X)$ is called a profile of A and we denote it by $extA$. The famous Krein-Milman Theorem (see cf. [62]) states that

THEOREM 1 (Krein-Milman). *Any $A \in cc(X)$ can be represented as*

$$A = clco\{extA\}.$$

In euclidean spaces R^l the situation is more exact. Namely,

THEOREM 2 (Carathéodory, see cf. [54]). *Let $A \in N(R^l)$. Then each point of coA can be represented as a convex combination of at most $l+1$ points from A.*

For a bounded set $A \subset R^l$ the Carathéodory Theorem implies that $co(clA)$ is compact and thus

$$(1.9) \qquad\qquad co(clA) = clcoA.$$

Indeed. Obviously $co(clA) \subset clcoA$. From the other hand we have

$$A \subset co(clA) = \left\{ \sum_{i=0}^{l} \lambda_i a_i : \lambda_i \geq 0, \ \sum_{i=0}^{l} \lambda_i = 1 \ \text{and} \ a_i \in clA \right\}$$

and $co(clA)$ is convex. It is also compact, since it is the image of the compact set

$$W = \left\{ w = (\lambda_0, ..., \lambda_l, a_0, ..., a_l) : \lambda_i \geq 0, \ \sum_{i=0}^{l} \lambda_i = 1 \ \text{and} \ a_i \in clA \right\}$$

by the continuous function

$$f(w) = f(\lambda_0, ..., \lambda_l, a_0, ..., a_l) = \sum_{i=0}^{l} \lambda_i a_i.$$

Therefore

$$clcoA \subset co(clA)$$

what gives (1.9).

Additionally, we may prove that for $A \in b(R^l)$ the formula

$$co\{ext(clA)\} = clcoA$$

holds. The latter can be explain more precisely by so called lexicographical orders in R^l. By this we mean an order given by a convex cone Θ satisfying (1.6). Each such cone Θ can be equivalently characterized by the family Ξ consisting of all orthonormal basis in R^l. Namely, Θ has the following form

$$\Theta = \left\{ x : \begin{array}{c} \langle x, \mathbf{a}_1 \rangle = ... = \langle x, \mathbf{a}_i \rangle = 0 \\ \text{and} \\ \langle x, \mathbf{a}_{i+1} \rangle > 0 \ \text{for some} \ i = 1, ..., l-1 \end{array} \right\},$$

where $\mathcal{E} = \{\mathbf{a}_1, \mathbf{a}_2, .., \mathbf{a}_l\}$ is an orthonormal basis in R^l. In such situation the lexicographical maximum $\max(W, \Theta)$ we shall also denote by $e(W, \mathcal{E})$. The motivation for an use of the letter e comes from the following

THEOREM 3. *For $A \in c(R^l)$ and any orthonormal basis $\mathcal{E} \in \Xi$ the lexicographical maximum $e(A, \mathcal{E})$ is an extreme point of A. And vice versa, for any extreme point $e \in A$ there is an orthonormal basis $\mathcal{E}$ such that*

$$e = e(A, \mathcal{E}).$$

PROOF. Assume first that $e = e(A, \mathcal{E})$ for some orthonormal basis $\mathcal{E} = \{\mathbf{a}_1, \mathbf{a}_2, .., \mathbf{a}_l\}$. If e is not an extreme point then there are $a, b \in A$, both different from e, and $\lambda \in (0, 1)$ such that

$$e = \lambda a + (1 - \lambda) b.$$

Since $a, b \underset{\mathcal{E}}{\leq} e = \lambda a + (1 - \lambda) b$ then there exist $i, k \in \{1, ..., l - 1\}$ such that for $j \in \{1, ..., i\}$

$$\langle a, \mathbf{a}_j \rangle = \langle \lambda a + (1 - \lambda) b, \mathbf{a}_j \rangle \text{ and } \langle a, \mathbf{a}_{i+1} \rangle < \langle \lambda a + (1 - \lambda) b, \mathbf{a}_{i+1} \rangle$$

and for $m \in \{1, ..., k\}$

$$\langle b, \mathbf{a}_m \rangle = \langle \lambda a + (1 - \lambda) b, \mathbf{a}_m \rangle \text{ and } \langle b, \mathbf{a}_{k+1} \rangle < \langle \lambda a + (1 - \lambda) b, \mathbf{a}_{k+1} \rangle.$$

Since $\lambda \in (0, 1)$, then, for $j \in \{1, ..., i\}$ and $m \in \{1, ..., k\}$, we have

$$\langle a, \mathbf{a}_j \rangle = \langle b, \mathbf{a}_j \rangle \text{ and } \langle a, \mathbf{a}_{i+1} \rangle < \langle b, \mathbf{a}_{i+1} \rangle$$

and

$$\langle b, \mathbf{a}_m \rangle = \langle a, \mathbf{a}_m \rangle \text{ and } \langle b, \mathbf{a}_{k+1} \rangle < \langle a, \mathbf{a}_{k+1} \rangle.$$

But this is impossible for $j = m = \min\{i, k\}$.

Let now $e \in A$ be an extreme point. Consider the space $E = span A$ with $dim E = k \leq l$ and take it's orthogonal complement

$$E^\perp = \{x : \langle x, a \rangle = 0 \text{ for every } a \in E\}.$$

If $dim E = l$ then there exists a unit vector $\mathbf{a}_1$ such that $\langle e, \mathbf{a}_1 \rangle = c_A(\mathbf{a}_1)$. Completing system $\{\mathbf{a}_1\}$ to an orthonormal basis in R^l one can check that this is a proper choice. If $dim E^\perp = l - k < l$ so we can pick an orthonormal basis $\{\mathbf{a}_1, \mathbf{a}_2, .., \mathbf{a}_{l-k}\}$ of $E^\perp$. Moreover $A \subset (E^\perp)^\perp$, so there is a unit vector $\mathbf{a}_{l-k+1} \in (E^\perp)^\perp$ such that

$$\langle e, \mathbf{a}_{l-k+1} \rangle = c_A(\mathbf{a}_{l-k+1}).$$

Now if we complete the system $\{\{\mathbf{a}_1, \mathbf{a}_2, .., \mathbf{a}_{l-k}, \mathbf{a}_{l-k+1}\}\}$ to an orthonormal basis $\mathcal{E}$ in R^l we can conclude that $e = e(A, \mathcal{E})$, what completes the proof. $\square$

REMARK 1. *Actually the above theorem determines that for $A \in c\left(R^l\right)$ the lexicographical maximum $e\left(A, \mathcal{E}\right)$ exists. In general this may not take place.*

The lexicographical order is very useful tool in convex analysis in R^l. It can be used for example in separation results. Namely, for two disjoint convex sets $W, U \subset R^l$ there is a basis $\mathcal{E} \in \Xi$ such that for any $(w, u) \in W \times U$ we have

$$w \underset{\mathcal{E}}{\leq} u.$$

The above remarks allow us describe the convex hull and the profile of any $W \in c\left(R^l\right)$.

THEOREM 4. *Let $W \in c\left(R^l\right)$. Then*

$$coW = clcoW = \bigcap_{\mathcal{E} \in \Xi}\left\{x : a \underset{\mathcal{E}}{\leq} e\left(W, \mathcal{E}\right)\right\}.$$

Moreover,

$$\{e\left(W, \mathcal{E}\right) : \mathcal{E} \in \Xi\} = ext\{coW\}.$$

PROOF. Denote by

$$W_0 = \bigcap_{\mathcal{E} \in \Xi}\left\{x : a \underset{\mathcal{E}}{\leq} e\left(W, \mathcal{E}\right)\right\}$$

and observe that W_0 is convex. Hence $coW \subset W_0$. To see that both sets coincide assume, to a contrary, that there is $p \in W_0 \backslash coW$. Therefore sets $\{p\}$ and coW can be separated by a basis $\mathcal{E} \in \Xi$, i.e.

$$e\left(coW, \mathcal{E}\right) \underset{\mathcal{E}}{<} p.$$

But this means that $p \notin W_0$, a contradiction.

It remains to show that

$$coW = clcoW.$$

Notice that for that it is sufficient check that for given $A \in b\left(R^l\right)$ any extreme point of $clcoA$ has to be in clA. For this purpose recall that by (1.9) we have

$$clcoA = co\left(clA\right) = \left\{\sum_{i=0}^{l} \lambda_i a_i : \lambda_i \geq 0, \quad \sum_{i=0}^{l} \lambda_i = 1 \quad and \quad a_i \in clA\right\}.$$

Fix now an extreme point $e \in clcoA$. By the above formula

$$e = \lim_{k \longrightarrow \infty} e_k,$$

where

$$e_k = \sum_{i=0}^{l} \lambda_{ik} a_{ik},$$

for some $\lambda_{ik} \geq 0$, $\sum_{i=0}^{l} \lambda_{ik} = 1$ *and* $a_{ik} \in A$. For every $i = 0, 1, ..., l$, passing to subsequences, if necessary, we may assume that $\lambda_{ik} \longrightarrow \lambda_i \geq 0$, $a_{ik} \longrightarrow a_i \in clA$ *and* $\sum_{i=0}^{l} \lambda_i = 1$.

Hence

$$e = \sum_{i=0}^{l} \lambda_i a_i.$$

But the extremality of e means that

$$e = a_0 = a_1 = ... = e_l \in clA,$$

what completes the proof.   □

The profile $extW$ of a given convex and compact set may not be closed. But complementarily to the provided results we have the following:

THEOREM 5. *Let* $W \in cc\left(R^l\right)$ *be given. Then each vector* $w \in W$ *can be represented as*

$$w = \sum_{i=0}^{l} \lambda_i e_i,$$

with $e_i \in extW$ *and* $\lambda_i \geq 0$ *such that* $\sum_{i=0}^{l} \lambda_i = 1$.

PROOF. We shall proceed by the induction.

For $l = 1$ it is clear since $W = [a, b]$.

Assume that the theorem holds for all $k \leq l-1$ and let $W \in cc\left(R^l\right)$ be given. Without loss of generality we may require that

$$e_l = 0 \in extW.$$

Fix $w \in W$. Since for $w = e_l$ the statement is clear then we need to cover just the case.

$$w \neq e_l = 0.$$

Consider the set

$$W_l = \{x \in W : \langle x, w \rangle = c_W(w)\}$$

and notice that W_l is convex and compact set with $\dim W_l \leq l - 1$. Moreover the vector

$$w_l = \frac{c_W(w)}{|w|^2} w \in W_l.$$

By the induction step there exist $e_i \in extW_l \subset extW$ and $\lambda_i \geq 0$ with $\sum_{i=0}^{l-1} \lambda_i = 1$ such that

$$w_l = \sum_{i=0}^{l-1} \lambda_i e_i.$$

Then

$$w = \sum_{i=0}^{l-1} \frac{\lambda_i |w|^2}{c_W(w)} e_i = \sum_{i=0}^{l} \widetilde{\lambda}_i e_i,$$

where

$$\widetilde{\lambda}_i = \frac{\lambda_i |w|^2}{c_W(w)} \quad i = 0, 1, ..., l - 1$$

and

$$\widetilde{\lambda}_l = 1 - \frac{|w|^2}{c_W(w)}.$$

Now an observation that all $\widetilde{\lambda}_i \geq 0$ ends the proof. $\qquad\square$

3. Measurable functions

Let T be arbitrary set with a $\sigma - field$ Σ of subset of T. Elements Σ we call of $\Sigma - measurable$ or, simply $measurable$, sets. As a model the reader may think on:

(1) the interval $I = [0, 1]$ and the real line R, both with a $\sigma - field$ $\mathcal{L}_0$ of Lebesgue measurable sets given by Lebesgue measure $\ell = dt$.

(2) a topological space T with a $\sigma - field$ $\mathcal{B} = \mathcal{B}(T)$ of Borel measurable sets, i.e. the $\sigma - field$ generated by open sets in T.

(3) a locally compact Hausdorff space T with a $\sigma - field$ $\mathcal{L}$ of Lebesgue measurable sets given by a nonnegative $\sigma-$ finite Radon measure μ;

(4) a complete metric space with a $\sigma - field$ $\mathcal{L}$ of Lebesgue measurable sets given by a locally finite Radon measure μ;

We can also consider product $\Sigma_1 \otimes \Sigma_2$ of $\sigma - fields$ Σ_1 and Σ_2 defined on T_1 and T_2, respectively. By this object it is usually meant the smallest $\sigma - field$ generated by sets $A_1 \times A_2$ with $A_i \in \Sigma_i$, $i = 1, 2$.

A function f from T into a topological space X is said to be measurable iff for every open set $U \subset X$ the set

(1.10) $$f^{-1}(U) = \{t : f(t) \in U\} \in \Sigma$$

As an example of measurable functions we deliver, so called, simple functions. By this term we mean a function $f : T \longrightarrow X$ assuming finite number of values $\{x_1, ..., x_n\}$, each of them on a measurable set, say x_i on $A_i \in \Sigma$. In other words

$$f = \begin{cases} x_1 \ for \ t \in A_1 \\ \quad .... \\ x_n \ for \ t \in A_n \end{cases} = \sum_{i=1}^{n} \chi_{A_i} x_i.$$

Evidently any simple simple function is measurable.

For metrizable topological spaces (X, d) the measurability of a given $f : T \longrightarrow X$ can be characterized in terms of real functions. Namely we have the following

PROPOSITION 5. *A function $f : T \longrightarrow X$ is measurable iff for every $x \in X$ the distance function*

$$t \longrightarrow d(x, f(t)) \ \ is \ \ measurable.$$

Moreover, if X is separable then is enough to check the above condition only for a dense subset in X.

As a conclusion from the above fact we have to emphasize that in arbitrary metric spaces it is enough to check the condition (1.10) just for open or closed balls. For real functions $f : T \longrightarrow R$ it is even simpler. Namely, it sufficies to verify that for each c one of the sets $\{t : f(t) > c\}$, $\{t : f(t) \geq c\}$, $\{t : f(t) < c\}$ or $\{t : f(t) \leq c\}$ is measurable.

Among real measurable functions an important role play upper and lower semicontinuous ones. Namely, we say that $f : T \longrightarrow R$ is:

lower semicontinuous (l.s.c. for short) iff for each c
the set $\{t : f(t) > c\}$ is open;
upper semicontinuous (u.s.c. for short) iff for each c
the set $\{t : f(t) < c\}$ is open.

Equivalently $f : T \longrightarrow R$ is:

l.s.c. iff for each c the set $\{t : f(t) \leq c\}$ is closed;
u.s.c. iff for each c the set $\{t : f(t) \geq c\}$ is closed.

Notice that $f : T \longrightarrow R$ is *l.s.c.* iff the function $-f$ is *u.s.c.*

The notion of lower semicontinuity can also be characterized with the use of epigraphs. By this term we mean a set

$$E(f) = epif = \{(t, c) : c \geq f(t)\}.$$

Then for $f : T \longrightarrow R$ the property of being an *l.s.c.* function is equivalent to closedness of the epigraph $E(f)$.

If $f : T_1 \times T_2 \longrightarrow X$ is $\Sigma_1 \otimes \Sigma_2 - measurable$ then we also say that it is jointly measurable.

The full details concerning the measurability properties are beyond the subject of this work and so we refer the reader monographies [55] and [190]. However, for our work to be self-contained we shall recall some basic properties and facts.

Unless there is not specifically stated we shall just assume that T is a complete separable metric space with a $\sigma - field$ $\mathcal{L}$ of Lebesgue measurable sets given by a locally finite Radon measure μ, while (X, d) stands for a separable metric space. Saying about measurability we mean $\mathcal{L} - measurability$.

For any measure space (S, Σ) we might be concerned with the projection

$$proj_T(A) = \{t \in T : (t, s) \in A \; for \; some \; s \in S\}$$

of a set $A \subset T \times S$ on T. Similarly we define the projection on S.

DEFINITION 2. *We shall say that (S, Σ) possess the projection property with respect to $(T, \mathcal{L}, \mu)$ iff for any $A \subset \mathcal{L} \otimes \Sigma$ we have*

$$(1.11) \qquad\qquad proj_T(A) \in \mathcal{L}.$$

As an example of such situation we provide the Borel $\sigma - field$ $\mathcal{B}$ of Borel measurable subsets of a topological space S.

THEOREM 6. *Let S be a topological space with $\sigma - field$ $\mathcal{B}$ of Borel measurable subsets of S. Then $\mathcal{B}$ possess the projection property with respect to $(T, \mathcal{L}, \mu)$.*

The results of this type are of main importance and were obtain by many authors. The presented formulation is due to M. Saint-Beuve (see also Castaing & Valadier [43]). We have to say that many results concerning joint a $\mathcal{L} \otimes \mathcal{B}-$measurability have their counterparts for abstract $\sigma - fields$ Σ if we postulate condition (1.11).

As a consequence of the Theorem 6 the reader may verify the following

PROPOSITION 6. *A function $f : T \longrightarrow R$ is $\mathcal{L}-measurable$ iff the epigraph $E(f) \in \mathcal{L} \times \mathcal{B}$.*

The space of all measurable functions from T into X we shall denote by $M(T, X)$, while $\mathcal{B} = \mathcal{B}(T, X)$ and $C = C(T, X)$ stand for the spaces, respectively, of Borel measurable functions and continuous functions. For $f, g \in M(T, X)$ we say that f and g are equal *almost everywhere* (*a.e.*) *in* T iff

$$\mu(\{t : f(t) \neq g(t)\}) = 0.$$

In such situations we identify f and g and write also $f = g$. Described identification is an equivalence relation and actually in $M(T, X)$ we have equivalence classes.

Introduced above spaces are metrizable by the formula (1.4), i.e.

$$\mathbf{d}(f, g) = \sup_{t \in T} \arctan d(f(t), g(t))$$

and they are complete whenever X is. Moreover, the inclusions

$$C(T, X) \subset \mathcal{B}(T, X) \subset M(T, X)$$

hold.

For a Banach space X all these spaces are locally convex and metrizable. If we restrict them to bounded functions

$$C_b(T, X) \subset \mathcal{B}_b(T, X) \subset M_b(T, X)$$

then they are Banach spaces with norm

(1.12)
$$\|f\|_\infty = \sup_{t \in T} |f(t)|$$

and the norm topologies are equivalent to the metric ones.

Let us also emphasize that in case of a Banach space X there is no need to distinguish between weak and strong measurability, since both notions coincide. Also with no lost of generality we may assume that X is separable. A reason for that provides a fact that any measurable function $f : T \longrightarrow X$ is *separable − valued*, i.e. then there exists a separable Banach space $X_f \subset X$ such that $f(t) \in X_f$ almost everywhere (*a.e.*) in T.

Measurable functions are close to continuous functions in the following sense:

THEOREM 7 (Lusin). *A function* $f : T \longrightarrow X$ *is measurable iff for every* $\varepsilon > 0$ *there exists a compact set* $T_\varepsilon \subset T$, *with* $\mu(T \backslash T_\varepsilon) \leq \varepsilon$ *such that* f *restricted to* K *is continuous.*

The Lusin Theorem has a generalization for sequence $\{f_n\}$ of measurable functions. Namely we have

THEOREM 8 (Yegorov). *Functions* $f_n \in M(T, X)$ *are almost everywhere convergent to* $f_0 \in M(T, X)$ *iff for every* $\varepsilon > 0$ *there exists a compact set* $T_\varepsilon \subset T$, *with* $\mu(T \backslash T_\varepsilon) \leq \varepsilon$, *such that* f_n *restricted to* T_ε *are uniformly convergent to* f_0. *Moreover, we may assume that all* f_n, $n = 0, 1, 2, ..$ *are continuous on* K.

3.1. Bochner Integrable Functions. By $L^p(T, X) = L^p(T, X, \mu)$ we mean the Banach space of Bochner integrable functions (equivalence classes) with the usual norm

$$\|u\|_p = \left(\int_T |u(t)|^p \, \mu(dt) \right)^{\frac{1}{p}} \quad for \ 1 \le p < \infty,$$

and

$$\|u\|_\infty = ess \sup |u(t)|.$$

If $X = R$ then use notation $L^p(T) = L^p(T, \mu) = L^p(T, R)$. The closed ball in $L^p(T, X)$ centered at u and with radius r we shall denote by $B_p(u, r)$.

A subset $K \subset M(T, X)$ is said to be $p - integrably$ bounded if there is an $a \in L^p(T, X)$ such that for every $u \in K$

$$|u(t)| \le a(t) \quad a.e. \ in \ T.$$

For $p = 1$ we simply say that K is integrably bounded.

For $p < \infty$ each $u \in L^p(T, X)$ can be obtained as the limit of simple function. This means that for every $\varepsilon > 0$ there is a simple function $u_\varepsilon \in L^p(T, X)$ such that

(1.13) $$\|u - u_\varepsilon\|_p < \varepsilon.$$

This observation leads to a very deep fact concerning integral operators. To explain this phenomenon let us notice that each $u \in L^p(T, X)$ produces a linear integral operator $L : L^q(T, R) \longrightarrow X$ given by

(1.14) $$Lh = \int_T h(t) u(t) \mu(dt),$$

where q is the conjugate exponent to p, i.e. $\frac{1}{p} + \frac{1}{q} = 1$.

Any integral operator L given by (1.14) is bounded with

$$\|L\| = \|u\|_p.$$

For a simple function $u = \sum_{i=1}^{n} x_i \chi_{A_i}$ with measurable A_i we have

$$Lh = \sum_{i=1}^{n} x_i \int_{A_i} h(t) \mu(dt)$$

and hence $L : L^q(T) \longrightarrow span\{x_1, ..., x_n\}$. Therefore L is finitely dimensional and thus compact.

For arbitrary $u \in L^p(T, X)$ the inequality (1.13) shows that for

$$L_\varepsilon h = \int_T h(t) u_\varepsilon(t) \mu(dt)$$

we have

$$\|L - L_\varepsilon\| = \|u - u_\varepsilon\|_p < \varepsilon.$$

The latter means that L given by (1.14) is the limit of compact operators L_ε and hence compact as well. So the arguments provided justify the following

THEOREM 9. *For every $u \in L^p(T, X)$, $p < \infty$, the operator $L : L^q(T, R) \longrightarrow X$ given by*

$$Lh = \int_T h(t) u(t) \mu(dt)$$

is compact.

Any measurable $A \in \mathcal{L}$ we identify with $\chi_A \in L^p(T)$ and in this manner $A = B$ means that $\chi_A = \chi_B$ in $L^p(T)$. Thus $\mathcal{L}$ endowded with

$$d_p(A, B) = \| \chi_A - \chi_B \|_p$$

is metrizable. But for every $p < \infty$

$$d_p(A, B) = \|\chi_A - \chi_B\|_p = [\mu(A \triangle B)]^{\frac{1}{p}},$$

where $A \triangle B$ stands for the symmetric difference of A and B. Therefore all these metrics are equivalent and hence we may identify $\mathcal{L}$ as a subset of $L^1(T)$.

PROPOSITION 7. *Within the above identification $\mathcal{L} \subset L^1(T)$ is closed and therefore complete.*

PROOF. To see the closedness of $\mathcal{L}$ let $\chi_{A_n} \longrightarrow u$ in $L^1(T)$. By the Riesz-Fischer Theorem we may extract a subsequence, still denoted by χ_{A_n}, such that χ_{A_n} is convergent a.e. to u. But each χ_{A_n} can assume the values from $\{0, 1\}$, so the same holds for u. Thus there exists $A \in \mathcal{L}$ such that $u = \chi_A$. □

REMARK 2. *For any given $0 \neq u \in L^p(T)$ the set*

$$U = \{u\chi_A : A \in \mathcal{L}\}$$

never is compact in $L^p(T)$. In particular, $\mathcal{L} \subset L^1(T)$ can not be compact. This can be explain later on, when we will have already discussed segments of nonatomic measures (see Proposition 7, Corollaries 3 and 4).

3.2. Mild functions. It what follows we shall assume that $(\mathcal{L}, d_1)$ is separable. In such case we say that the measure μ is separable. Moreover, the separability of μ implies in our case that also the spaces $L^p(T, X)$ are separable for $p < \infty$.

Denote by $\mathcal{I}(\zeta, v) : X \times L^1(I, X) \longrightarrow C(I, X)$ an operator given by

$$\{\mathcal{I}(\zeta, v)\}(t) = \zeta + \int_0^t v(\tau)\, d\tau.$$

A function $u : I \longrightarrow X$ we shall call mild or absolutely continuous iff there exists an integrable $v \in L^1(I, X)$ such that u can be represented in the form

$$u(t) = \mathcal{I}\{u(0), v\} = u(0) + \int_0^t v(\tau)\, d\tau, \quad t \in I.$$

The function v plays a role of the derivative and sometimes we shall denote it by

$$v = \frac{du}{dt} = u'.$$

But we should be aware that the formula

$$(1.15) \quad \left(\frac{d}{dt} \circ \mathcal{I}\right)(v) = \frac{d}{dt}\left(u_0 + \int_0^t v(\tau)\, d\tau\right) = v(t) \quad a.e. \;\; in \;\; I$$

holds only in Banach spaces X with the Radon-Nikodym Property (see cf. [55]), for example in finite dimensional or reflexive X. For such Banach spaces the notion of being a mild function is equivalent to the classical absolutely continuouity, that is:

For every $\varepsilon > 0$ there exists $\delta > 0$ such that for every finite sequence $\{I_k\}$ of disjoint intervals, $I_k = (a_k, b_k)$, such that $\sum (b_k - a_k) < \delta$ we have

$$\sum_k |u(b_k) - u(a_k)| < \varepsilon.$$

Mild functions form a Banach space with the norm

$$\|u\|_0 = |u(0)| + \|u'\|_1 = |u(0)| + \int_0^1 |u'(\tau)|\, d\tau.$$

This space we shall denote by $AC = AC(I, X)$. Notice that

$$AC(I, X) \subset C(I, X)$$

and
$$\|u\|_\infty = \sup_{t\in T} |u(t)| \leq \|u\|_0 .$$

We also need a connection between continuity of a mapping from a topological space S into $L^1(I,X)$ with continuity into $AC(I,X)$.

PROPOSITION 8. *Assume that* $v : S \longrightarrow L^1(I,X)$ *and* $\zeta : S \longrightarrow X$ *are continuous. Then the mapping*

$$s \longrightarrow \mathcal{I}\{\zeta(s), v(s)\}(t) = \zeta(s) + \int_0^t v(s)(\tau)\,d\tau$$

is continuous from S into $AC(I,X)$.

A proof we leave as an exercise.

We should point out that in the theory of differential equations or differential inclusions solutions which are absolutely continuous in the above sense are also called *mild solutions*.

CHAPTER 2

Real and vector measures

Let T be arbitrary set with a $\sigma - field$ Σ of subset of T. By a partition of a set $A \in \Sigma$ we understand a countable family $\{A_n\}_{n=1}^{\infty} \subset \Sigma$ of pairwise disjoint measurable sets such that $A = \bigcup\limits_{n=1}^{\infty} A_n$. If a partition $\{A_n\}_{n=1}^{\infty}$ consists only of finitely many nonvoid sets we say that it is a finite one.

DEFINITION 3. *A mapping* $m : \Sigma \longrightarrow X$ *we call a vector measure if it possess the following property:*
for any sequence $\{A_n\}_{n=1}^{\infty} \subset \Sigma$ *of pairwise disjoint measurable sets*

$$(2.1) \qquad m\left(\bigcup_{n=1}^{\infty} A_n\right) = \sum_{n=1}^{\infty} m(A_n),$$

where the series on the right hand-side is absolutely summable.
By the range of a vector measure $m : \Sigma \longrightarrow X$ *we mean the set*

$$\mathcal{R}(m) = \{m(A) : A \in \Sigma\}.$$

The family of all measures $m : \Sigma \longrightarrow X$ *we denote by* $\mathcal{M}(X) = \mathcal{M}(T, \Sigma, X)$.
If $X = R$ *then we shall denote* $\mathcal{M} = \mathcal{M}(T) = \mathcal{M}(T, \Sigma, R)$ *and any member* $m \in \mathcal{M}$ *call a real measure.*

The condition (2.1) we also equivalently express in terms of monotonic sequences $\{A_n\}_{n=1}^{\infty} \subset \Sigma$.

DEFINITION 4. *We say that sequence* $\{A_n\}_{n=1}^{\infty}$ *is:*

increasing iff $\quad A_1 \subset A_2 \subset ... \subset A_n \subset ...$
    and
decreasing iff $\quad A_1 \supset A_2 \supset ... \supset A_n \supset ...$

PROPOSITION 9. *Assume that a mapping* $m : \Sigma \longrightarrow X$ *posses the property: for any disjoint sets* $A, B \in \Sigma$ *we have*

$$m(A \cup B) = m(A) + m(B).$$

Then the following conditions are equivalent:

23

    i. m is a vector measure;

    ii. for every increasing family $\{A_n\}_{n=1}^{\infty} \subset \Sigma$ we have

$$m\left(A_n\right) \longrightarrow m\left(\bigcup_{n=1}^{\infty} A_n\right) \quad as \ n \longrightarrow \infty;$$

    iii. for every increasing family $\{A_n\}_{n=1}^{\infty} \subset \Sigma$ such $\bigcup_{n=1}^{\infty} A_n = T$ that we have

$$m\left(A_n\right) \longrightarrow m\left(T\right) \quad as \ n \longrightarrow \infty;$$

    iv. for every decreasing family $\{A_n\}_{n=1}^{\infty} \subset \Sigma$ there is

$$m\left(A_n\right) \longrightarrow m\left(\bigcap_{n=1}^{\infty} A_n\right) \quad as \ n \longrightarrow \infty.$$

    v. for every decreasing family $\{A_n\}_{n=1}^{\infty} \subset \Sigma$ such that $\bigcap_{n=1}^{\infty} A_n = \emptyset$ there is

$$m\left(A_n\right) \longrightarrow 0 \quad as \ n \longrightarrow \infty.$$

1. Real measures

In real case we can also allow a measure m to assume one of the improper values $+\infty$ or $-\infty$ (but not both). In such case we write $m : \Sigma \longrightarrow \overline{R}$, where by $\overline{R}$ we mean $R \cup \{+\infty\}$ or $R \cup \{-\infty\}$, and call m again real measure or extended real measure. The family of all extended real measures we shall again denote by $\mathcal{M}$.

For real measure $m \in \mathcal{M}$ we have the following:

PROPOSITION 10. *The values of a real measure m which never assume the value $+\infty$ are bounded from above.*

PROOF. For the purpose of the proof let us call a set $A_0 \in \Sigma$ to be unbounded if

$$\sup \{m\left(A \cap A_0\right) : A \in \Sigma\} = +\infty.$$

To a contrary let us assume that the measure m has no upper bound i.e. the set T is unbounded. Then the following two situations may occur:

    (1) every unbounded set contains a measurable unbounded set of arbitrarily large measure;

    (2) there is an unbounded set B_0 and an integer n_0 such that B_0 contains no unbounded set of measure grater then n_0.

In the case 1. one can inductively construct a decreasing family $\{A_n\}$ of unbounded sets such that $m\,(A_n) \geq n$. Then for $A_0 = \bigcap\limits_{n=1}^{\infty} A_n$ we have, by the Proposition 9

$$m\,(A_0) = m\left(\bigcap_{n=1}^{\infty} A_n\right) = \lim m\,(A_n) = +\infty,$$

what contradicts our hypothesis.

In the case 2., since $\sup\{m\,(A \cap B_0) : A \in \Sigma\} = +\infty$, there exists a measurable subset $B_1 \subset B_0$ such that $m\,(B_1) > n_0$. By our assumptions the set B_1 is not unbounded. Hence $B_0 \backslash B_1$ is unbounded, because otherwise

$$\sup\{m\,(A \cap B_0) : A \in \Sigma\} =$$
$$= \sup\{m\,(A \cap B_1) + m\,(A \cap (B_0 \backslash B_1)) : A \in \Sigma\} \leq$$
$$\leq \sup\{m\,(A \cap B_1) : A \in \Sigma\} + \sup\{m\,(A \cap (B_0 \backslash B_1)) : A \in \Sigma\} < +\infty.$$

Therefore there exists a measurable $A_1 \subset B_0 \backslash B_1$ with $m\,(A_1) \geq 1$. Take $B_2 = A_1 \cup B_1$ and observe that $B_2 \subset B_0$ with $m\,(B_2) > n_0$. Again B_2 is not unbounded and hence $B_0 \backslash B_2$ is unbounded. Continuing this procedure we may pick up a sequence of disjoint measurable sets $\{A_n\}$ with the property that $m\,(A_n) \geq 1$ that for all n. But then for $A_0 = \bigcup\limits_{n=1}^{\infty} A_n$ we have

$$m\,(A_0) = m\left(\bigcup_{n=1}^{\infty} A_n\right) = \sum_{n=1}^{\infty} m\,(A_n) = \infty,$$

a contradiction. □

DEFINITION 5. *A triple* (T, Σ, m), *where* m *is a real measure we shall call a measure space. The measure space* (T, Σ, m) *is finite if* $m : \Sigma \longrightarrow R$.

Note that for a finite measure space (T, Σ, m) the range $\mathcal{R}\,(m)$ is, by the Proposition 10, bounded. This in particular means that for such measure

$$\sup\{m\,(A) : A \in \mathcal{L}\} < \infty.$$

For $m \in \mathcal{M}$ by m^+ and m^- we denote positive and negative parts of m i.e.

$$m^+\,(A) = \sup\{m\,(B) : B \subset A, \quad B \in \mathcal{L}\}$$

and

$$m^-\,(A) = \sup\{-m\,(B) : B \subset A, \quad B \in \mathcal{L}\}.$$

Recall that (see [**62**])

$$m^+, m^- \in \mathcal{M}.$$

Moreover, we have the Hahn decomposition

$$m = m^+ - m^-$$

and total variation

$$|m| = m^+ + m^-.$$

The space $\mathcal{M}$, normed by $\|m\| = |m|(T)$, is a Banach space. It is also an ordered Banach lattice. Namely, for $m, \nu \in \mathcal{M}$ we say that $m \le \nu$ if for every $A \in \mathcal{L}$ the inequality $m(A) \le \nu(A)$ holds. If $0 \le m$ then we say that m is nonnegative. All nonnegative measures we denote by $\mathcal{M}^+$.

As an easy conclusion from the definition of positive and negative parts we can derive the following

PROPOSITION 11. *For $m, \nu \in \mathcal{M}$ the following conditions are equivalent:*

i. $m \le \nu$;
ii. $m^+ \le \nu^+$ and $m^- \ge \nu^-$.

2. Vector measures

Denote by $\mathcal{M}(X) = \mathcal{M}(T, \Sigma, X)$ the family of all measures $m : \Sigma \longrightarrow X$. We begin with the following

PROPOSITION 12. *The range $\mathcal{R}(m)$ of a vector measure $m \in \mathcal{M}(X)$ is a bounded set.*

PROOF. Fix $x^* \in X^*$ and denote by $\mu(A) = \langle x^*, m(A) \rangle$. One can easily check that $\mu : \Sigma \longrightarrow R$ is a real-valued measure. Therefore its range $\mathcal{R}(\mu)$ is bounded. But the latter means that for every $x^* \in X^*$ we have

$$\sup\{|x| : x \in \mathcal{R}(\mu)\} = \sup\{|\langle x^*, m(A)\rangle| : A \in \Sigma\} < \infty.$$

Now an easy application of the Banach-Steinhaus Boundedness Principle ends the proof.   □

DEFINITION 6. *By the total variation of a vector measure $m : \Sigma \longrightarrow X$ on a set $A \in \Sigma$ it is usually meant*

$$|m|(A) = \sup\left\{\sum_{k=1}^{n} |m(A_k)| : \{A_k\}_{k=1}^{n} \text{ is a finite partition of } A\right\} =$$

$$= \sup\left\{\sum_{k=1}^{n} |m(A_k \cap A)| : \{A_k\}_{k=1}^{n} \text{ is a finite partition of } T\right\}.$$

PROPOSITION 13. *The triple $(T, \Sigma, |m|)$, where $|m|$ is the total variation of a vector measure $m : \Sigma \longrightarrow X$, is a finite measure space. This in particular means that $|m|(T)$ is finite.*

PROOF. Take arbitrarily a set $A \in \Sigma$ and a finite partition $\{A_k\}_{k=1}^n$ of A. By the Hahn-Banach Theorem there exist $x_k^* \in X^*$, $k \in \{1, ..., n\}$, such that

$$|x_k^*| = 1 \quad and \quad |m(A_k)| = \langle x_k^*, m(A_k) \rangle.$$

Denote by

$$N_+ = \{k \in \{1, ..., n\} : \langle x_k^*, m(A_k) \rangle \geq 0\}$$

and by

$$N_- = \{k \in \{1, ..., n\} : \langle x_k^*, m(A_k) \rangle < 0\}.$$

Then

$$\sum_{k=1}^n |\langle x^*, m(A_k) \rangle| = \sum_{k \in N_+} \langle x^*, m(A_k) \rangle - \sum_{k \in N_-} \langle x^*, m(A_k) \rangle =$$

$$= \left\langle x^*, m \left(\bigcup_{k \in N_+} A_k \right) \right\rangle - \left\langle x^*, m \left(\bigcup_{k \in N_-} A_k \right) \right\rangle$$

$$\leq 2 |x^*| \sup \{|x| : x \subset \mathcal{R}(m)\}$$

and hence we have

$$\sum_{k=1}^n |m(A_k)| = \sup \left\{ \sum_{k=1}^n |\langle x^*, m(A_k) \rangle| : |x^*| \leq 1 \right\} \leq$$

$$\leq 2 \sup \{|x| : x \in \mathcal{R}(m)\}.$$

Thus

$$|m|(A) \leq 2 \sup \{|x| : x \in \mathcal{R}(m)\},$$

what gives the required boundedness. $\square$

Moreover m is absolutely continuous with respect to its total variation. Thus, assuming the Banach space X has the Radon-Nikodym property, m can be uniquely represented in the form

$$(2.2) \qquad m(A) = \int_A f(t) |m|(dt),$$

with density function $f \in L^1(T, X, |m|)$. Additionally, $|f(t)| = 1$ a.e. with respect to $|m|$. Since every euclidean space R^l admits the Radon-Nikodym property, the representation (2.2) holds for any nonatomic vector measure $m : \Sigma \longrightarrow R^l$.

In further considerations let us fix a nonnegative measure $\mu \in M(T, \Sigma)$.

A subset of $\mathcal{M}(T, X)$ consisting of those measures m which are given by a density function $f = \frac{dm}{d\mu} \in L^1(T, X, \mu)$ we shall denote by $\mathcal{M}_a(X)$.

For a measure $m : \Sigma \longrightarrow X$ by the restriction of m to a set $B \in \Sigma$ we mean the measure

$$m_{|B}(A) = m(A \cap B).$$

PROPOSITION 14.   (i) *For any* $m \in \mathcal{M}_a(X)$ *we have*

$$\|m\| = \left\| \frac{dm}{d\mu} \right\|_1.$$

(ii) *Let S be a topological space. Then a mapping $m = m_s$ from S into $\mathcal{M}_a(X)$ is continuous iff the mapping $\frac{dm_s}{d\mu} : S \longrightarrow L^1(T, X, \mu)$ is continuous.*

(iii) *Recall that by the Vitaly-Hahn-Saks Theorem, see cf. [62], for any continuous $m : S \longrightarrow \mathcal{M}_a(T, X)$ and $A : S \longrightarrow \Sigma$ the mapping $\widetilde{m} : S \longrightarrow \mathcal{M}_a(T, X)$ given by $\widetilde{m}(s) = m(s)_{|A(s)}$ is continuous.*

Sometimes there is a need to compare measures $m_k \in \mathcal{M}(T_k, \Sigma_k, X_k)$, $k = 1, 2$, provided that spaces X_1 and X_2 are isometric.

DEFINITION 7. *We say that measures $m_k \in \mathcal{M}(T_k, \Sigma_k, X_k)$, $k = 1, 2$ are isomorphic iff there is a mapping $\tau : \Sigma_1 \overset{onto}{\longrightarrow} \Sigma_2$ such that*

i.   *for any sequence $\{A_n\}_{n=1}^{\infty} \subset \Sigma_1$ of pairwise disjoint sets we have*

$$\tau\left(\bigcup_{n=1}^{\infty} A_n \right) = \bigcup_{n=1}^{\infty} \tau(A_n);$$

ii.   *for every $A \in \Sigma_1$*

$$m_2\{\tau(A)\} = m_1(A).$$

Equivalently i. it can be formulated in the following way:

i'. *for any decreasing sequence $\{A_n\}_{n=1}^{\infty} \subset \Sigma_1$ such that $A = \bigcap_{n=1}^{\infty} A_n$ the relation*

$$\tau(A) = \bigcap_{n=1}^{\infty} \tau(A_n)$$

holds.

For example measures $m_1 : \Sigma \longrightarrow R$ and $m_2 = m_1 x : \Sigma \longrightarrow X$, where $0 \neq x \in X$ is a given vector, are isomorphic. In general (see [133])

THEOREM 10. *Nonatomic measures $m_k \in \mathcal{M}(T_k, \Sigma_k, X_k)$, $k = 1, 2$ are isomorphic iff any isometry of X_1 and X_2 can be extended to an isometry of $L^1(T_1, \Sigma_1, X_1, m_1)$ and $L^1(T_2, \Sigma_2, X_2, m_2)$.*

3. Essential suprema and infima

3.1. Essential supremum and infimum of a family of real measures.

Let Λ be an index set and consider a family

$$\{m_\lambda\}_{\lambda \in \Lambda} \subset \mathcal{M} = \mathcal{M}(T, \overline{R}).$$

DEFINITION 8. *We say that the family $\{m_\lambda\}_{\lambda \in \Lambda}$ is:*

   i. *bounded from below by a measure $\mu_0 \in \mathcal{M}$ if for every $\lambda \in \Lambda$ we have*

(2.3) $$m_\lambda \geq \mu_0;$$

   ii. *bounded from above by a measure $\mu_0 \in \mathcal{M}$ if for every $\lambda \in \Lambda$ we have*

(2.4) $$m_\lambda \leq \mu_0;$$

   iii. *bounded by a measure $\mu_0 \in \mathcal{M}$ if for every $\lambda \in \Lambda$ we have* $-\mu_0 \leq m_\lambda \leq \mu_0$.

Any measure μ_0 satisfying (2.3) {(2.4)} is called a lower {upper} bound (measure) for $\{m_\lambda\}_{\lambda \in \Lambda}$.

Of course if a family $\{m_\lambda\}_{\lambda \in \Lambda}$ is bounded from below by a measure $\mu_0 \in \mathcal{M}$ then the family $\{-m_\lambda\}_{\lambda \in \Lambda}$ is bounded from above by a measure $-\mu_0$. Moreover, any bounded family of measures is bounded from above and from below.

We are going to discuss the following questions:

   (1) does any bounded from below {above} family of measures admit the smallest lower {the greatest upper} bound and in what sense?

   (2) does the boundedness from above and from below imply the boundedness?

DEFINITION 9. *Let $\{m_\lambda\}_{\lambda \in \Lambda} \subset \mathcal{M}$ be a given family of measures. A measure $m \in \mathcal{M}$ is called the infimum (supremum) measure for $\{m_\lambda\}_{\lambda \in \Lambda}$ if it satisfies the following conditions:*

   i. *m is an lower (upper) bound for $\{m_\lambda\}_{\lambda \in \Lambda}$;*

   ii. *for any lower (upper) bound v for $\{m_\lambda\}_{\lambda \in \Lambda}$ we have $m \geq v$ $(m \leq v)$.*

The infimum measure for $\{m_\lambda\}_{\lambda \in \Lambda}$ we shall denote by $\inf_{\lambda \in \Lambda} \{m\lambda\}$, while the supremum measure by $\sup_{\lambda \in \Lambda} \{m_\lambda\}$. If the index set is finite then *the infimum measure for* $\{m_\lambda\}_{\lambda\in\Lambda}$ *we shall call the minimum measure for* $\{m_\lambda\}_{\lambda\in\Lambda}$ *and denote by* $\min\limits_{\lambda\in\Lambda}\{m_\lambda\}$. *Similarly in such situation the supremum measure we shall call the maximum measure for* $\{m_\lambda\}_{\lambda\in\Lambda}$ *and denote by* $\max\{m_\lambda\}$.

An easy conclusion from the above definition is the following:

PROPOSITION 15. *Let* $\{m_\lambda\}_{\lambda\in\Lambda}\subset\mathcal{M}$ *be a given family of measures. Then*

   i. $m=\inf\limits_{\lambda\in\Lambda}\{m_\lambda\}$ *if and only if* $-m=\sup\limits_{\lambda\in\Lambda}\{-m_\lambda\}$

   ii.   *for any* $m_0\in\mathcal{M}$ *we have* $\inf\limits_{\lambda\in\Lambda}\{m_\lambda+m_0\}=\inf\limits_{\lambda\in\Lambda}\{m_\lambda\}+m_0$;

$$iii.\quad \inf\limits_{\lambda\in\Lambda}\{cm_\lambda\}=\begin{cases} c\inf\limits_{\lambda\in\Lambda}\{m_\lambda\} & for \quad c\geq 0; \\ c\sup\limits_{\lambda\in\Lambda}\{m_\lambda\} & for \quad c<0. \end{cases}$$

We leave the proof for the reader.

We need to point out that, by the Proposition 15, we may restrict our considerations to the existence of infimum measure. Moreover, an examination of a family $\{m_\lambda\}_{\lambda\in\Lambda}\subset\mathcal{M}$ bounded from below by measure m we can always reduce to the case of $\{m_\lambda\}_{\lambda\in\Lambda}\subset\mathcal{M}^+$, since $m_\lambda\geq m\Longleftrightarrow m_\lambda-m\geq 0$.

A situation when we have positive answer for the first question describes the following:

THEOREM 11. *Let* $\{m_\lambda\}_{\lambda\in\Lambda}\subset\mathcal{M}^+$ *be a given family of measures. Then there exists an infimum measure. If, additionally, a family* $\{m_\lambda\}_{\lambda\in\Lambda}\subset\mathcal{M}$ *is bounded from above by a separable measure* $\mu_0\in\mathcal{M}$ *then one can pick up a countable family* $\{m_{\lambda_n}\}_{n=1}^\infty$ *such that*

$$\inf\limits_{\lambda\in\Lambda}\{m_\lambda\}=\inf\limits_{n}m_{\lambda_n}.$$

PROOF. Define a mapping $m:\mathcal{L}\longrightarrow R$ by

$$m(A)=\inf\left\{\sum_i m_{\lambda_i}(C_i\cap A): \begin{array}{l} \{\lambda_i\}\subset\Lambda \ \ is \ \ finite, \\ \{C_i\}\subset\mathcal{L} \ \ is \ \ a \ \ finite \ \ partition \ \ of \ T \end{array}\right\}.$$

Clearly, for every all $\alpha\in\Lambda$ and $A\in\mathcal{L}$ we have

$$(2.5)\quad m(A)=\inf\left\{\sum_i m_{\lambda_i}(C_i\cap A)\right\}\leq m_{\lambda_0}(\emptyset)+m_\lambda(A)=m_\lambda(A)$$

and hence the condition *i.* in the definition 9 holds. We shall show that it is the required infimum measure. For this purpose we invoke the Theorem 9v.

Let A, B be two disjoint members of $\mathcal{L}$. We need to verify that

(2.6) $$m(A \cup B) = m(A) + m(B).$$

Fix $\varepsilon > 0$. Choose finite partitions $\{C_i\} \subset \mathcal{L}$ and $\{D_j\} \subset \mathcal{L}$ of T satisfying for some $\{\lambda_i\} \subset \Lambda$ and $\{\beta_j\} \subset \Lambda$ conditions

$$\sum_i m_{\lambda_i}(C_i \cap A) \leq m(A) + \varepsilon \quad \text{and} \quad \sum_j m_{\beta_j}(D_j \cap B) \leq m(B) + \varepsilon.$$

Then the collection $\{C_i \cap D_j\}$ is such finite partition of T that

$$m(A \cup B) \leq \sum_{i,j} m_{\lambda_i}(C_i \cap D_j \cap A) + \sum_{i,j} m_{\beta_j}(C_i \cap D_j \cap B) =$$

$$= \sum_i m_{\lambda_i}(C_i \cap A) + \sum_j m_{\beta_j}(D_j \cap B) \leq m(A) + m(B) + 2\varepsilon.$$

Thus

(2.7) $$m(A \cup B) \leq m(A) + m(B).$$

On the other hand we can select a finite partition $\{C_i\} \subset \mathcal{L}$ of $A \cup B$ and finite $\{\lambda_i\} \subset \Lambda$ such that

$$\sum_i m_{\lambda_i}(C_i \cap (A \cup B)) \leq m(A \cup B) + \varepsilon.$$

Then $\{C_i \cap A\}$ and $\{C_i \cap B\}$ are, respectively, finite partitions of A and B such that

$$m(A) + m(B) \leq \sum_i m_{\lambda_i}(C_i \cap A) + \sum_i m_{\lambda_i}(C_i \cap B) =$$

$$= \sum_i m_{\lambda_i}(C_i \cap (A \cup B)) \leq m(A \cup B) + \varepsilon.$$

Therefore

$$m(A) + m(B) \leq m(A \cup B),$$

what together with (2.7) justifies (2.6). To the end of the proof that m is a measure choose a decreasing family $\{A_n\} \subset \mathcal{L}$ with void intersection. Then, by (2.5),

$$0 \leq m(A_n) \leq m_\lambda(A_n),$$

what implies that $m(A_n) \longrightarrow 0$, as desired.

To see that the condition ii in 9 holds take any $v \in \mathcal{M}^+$ such that, for every $\lambda \in \Lambda$, we have $m_\lambda \geq v$. Similarly, as for m, we may conclude that for every finite $\{\lambda_i\} \subset \Lambda$, arbitrary finite partition $\{A_i\} \subset \mathcal{L}$ of T and any $A \in \mathcal{L}$ an inequality

$$\sum_i m_{\lambda_i}(A_i \cap A) \geq \sum_i v(A_i \cap A) = v(A)$$

holds. Hence

$$m(A) = \inf \left\{ \sum_i m_{\lambda_i}(A_i \cap A) \right\} \geq v(A).$$

Consider now the case that there is a separable measure $\mu_0 \in \mathcal{M}$ such that for every $\lambda \in \Lambda$

$$0 \leq m_\lambda \leq m_0.$$

Recall that in this case the separability means that the metric space $(\mathcal{L}, \mu_0)$ is separable.

Let $\{A_n\}_{n=1}^{\infty}$ be a dense subset in $(\mathcal{L}, \mu_0)$. For every $n, i = 1, 2, \ldots$ one can pick up finite subsets $\Lambda_{i,n} \subset \Lambda$ and finite partitions

$$\Pi_{i,n} = \{A_\lambda : \lambda \in \Lambda_{i,n}\}$$

of T such that

$$m(A_n) = \inf_i \left\{ \sum_{\lambda \in \Lambda_{i,n}} m_\lambda(A_\lambda \cap A_n) \right\}.$$

Define $\Lambda_0 = \bigcup_{i,n=1}^{\infty} \Lambda_{i,n}$ and notice that it is countable. Take

$$v_0 = \inf_{\lambda \in \Lambda_0} \{m_\lambda\}.$$

We shall prove that

$$m = \inf_{\lambda \in \Lambda} \{m_\lambda\} = \inf_{\lambda \in \Lambda_0} \{m_\lambda\} = v_0.$$

Obviously we have

$$0 \leq m \leq v_0 \leq \mu_0.$$

For the opposite inequality $m \geq v_0$ observe that for every $\lambda \in \Lambda_0$ and $A \in \mathcal{L}$ we have

(2.8) $$m_\lambda(A) \geq v_0(A).$$

Hence for every given $n, i = 1, 2, \ldots$

$$\sum_{\lambda \in \Lambda_{i,n}} m_\lambda(A_\lambda \cap A_n) \geq \sum_{\lambda \in \Lambda_{i,n}} v_0(A_\lambda \cap A_n) = v_0(A_n).$$

Thus

$$m\left(A_n\right) = \inf_i \left\{ \sum_{\lambda \in \Lambda_{i,n}} m_\lambda \left(A_\lambda \cap A_n\right) \right\} \geq v_0\left(A_n\right).$$

The latter, together with (2.8), stands for that the inequality

$$(m - v_0)\left(A\right) \geq 0$$

holds for every member $A \subset \{A_n\}_{n=1}^\infty$. But

$$0 \leq v_0 - m \leq \mu_0$$

and therefore the measure $v_0 - m$ is absolutely continuous with respect to μ_0. Hence the separability of μ_0 implies that $v_0 - m$ is separable as well and therefore $\{A_n\}_{n=1}^\infty$ is a dense subset in $(\mathcal{L}, v_0 - m)$. This in turn gives that $m = v_0$, what completes the proof. $\qquad\square$

COROLLARY 1. *If a family* $\{m_\lambda\}_{\lambda \in \Lambda} \subset \mathcal{M}$ *is bounded from below by* m_b *and from above by* m_a *then it is bounded by the measure*

$$m_0 = \max\{m_a, -m_b\}.$$

Checking this we leave the reader as an exercise.

3.2. Essential supremum and infimum of a family of functions. The previous consideration are a basis of the notion of essential supremum and infimum of an arbitrary family $\{u_\lambda\}_{\lambda \in \Lambda} \subset \mathcal{M}\left(T, \overline{R}\right)$ of measurable functions. Recall that for arbitrary $u, v \in \mathcal{M}\left(T, \overline{R}\right)$ the symbol $u \leq v$ stands for that $u\left(t\right) \leq v\left(t\right)$ *a.e.* in T.

We begin with

DEFINITION 10. *A measurable function* $u : T \longrightarrow \overline{R}$ *is said to be the essential infimum (essential supremum) of the family* $\{u_\lambda\}_{\alpha \in \Lambda} \subset \mathcal{M}\left(T, \overline{R}\right)$ *if*

   i. *for every* $\lambda \in \Lambda$ *there is* $u_\lambda \geq u$;
   ii. *if for some* $v : T \longrightarrow \overline{R}$ *and for every* $\lambda \in \Lambda$ *the inequality* $u_\lambda \geq u$ $(u_\lambda \leq u)$ *holds then* $u \geq v$ $(u \leq v)$.

The *essential infimum (essential supremum)* is denoted by

$$ess\inf\{u_\lambda\} \ (ess\sup\{u_\lambda\}) \quad or \quad ess\inf_{\lambda \in \Lambda}\{u_\lambda\} \left(ess\sup_{\lambda \in \Lambda}\{u_\lambda\}\right).$$

Before we prove that the essential infimum is always well-defined we give some it's properties.

PROPOSITION 16. *The essential infimum and essential supremum have the following useful properties:*
   i. $u = ess\inf\{u_\lambda\}$ *if and only if* $-u = ess\sup\{-u_\lambda\}$;

    ii.   for any $a \in \mathcal{M}\left(T, \overline{R}\right)$

$$\operatorname{ess\,sup}\left\{u_\lambda + a\right\} = \operatorname{ess\,sup}\left\{u_\lambda\right\} + a;$$

(2.9) $$\operatorname{ess\,inf}\left\{u_\lambda + a\right\} = \operatorname{ess\,inf}\left\{u_\lambda\right\} + a.$$

    iii.

$$\operatorname{ess\,inf}\left\{cu_\lambda\right\} = \begin{cases} c\,\operatorname{ess\,inf}\left\{u_\lambda\right\} & for \quad c \geq 0 \\ c\,\operatorname{ess\,sup}\left\{u_\lambda\right\} & for \quad c < 0 \end{cases}.$$

We leave the proof for the reader.

The above Proposition allow to concentrate just on essential infima. Having this in mind we can pass to the existence of prescribed above objects.

THEOREM 12. *For a given family $\left\{u_\lambda\right\}_{\lambda \in \Lambda} \subset \mathcal{M}\left(T, \overline{R}\right)$ there exist $\operatorname{ess\,inf}\left\{u_\lambda\right\}$. Moreover one can pick up a countable family $\left\{u_{\lambda_n}\right\}_{n=1}^{\infty}$ such that*

(2.10) $$\operatorname{ess\,inf}\left\{u_\lambda\right\} = \inf_n u_{\lambda_n}.$$

Similar statements hold for essential maxima.

PROOF. Our construction proceeds in two steps.
Step 1.   Suppose first that all u_λ are bounded by an integrable function, i.e. there exists $\varphi \in L^1\left(T, R\right)$ such that for every $\lambda \in \Lambda$

$$-\varphi \leq u_\lambda \leq \varphi$$

or, equivalently,

$$0 \leq u_\lambda + \varphi \leq 2\varphi.$$

Replacing $\left\{u_\lambda\right\}$ with $\left\{u_\lambda + \varphi\right\}$ we may assume, in view of the Proposition 16, that for a function $a \in L^1\left(T, R\right)$ such that

$$0 \leq u_\lambda \leq a.$$

Take measures m_λ and μ_0 with densities, respectively, u_λ and a, i.e. for every $A \in \left\{A_i\right\} \subset \mathcal{L}$ we have

$$m_\lambda\left(A\right) = \int_A u_\lambda\left(t\right) \mu\left(dt\right)$$

and

$$\mu_0\left(A\right) = \int_A a\left(t\right) \mu\left(dt\right).$$

By the Theorem 11 the measure

$$m = \inf m_\lambda$$

is well defined. One can easily check that

(2.11) $$0 \leq m \leq m_\lambda \leq \mu_0.$$

Since μ_0 is absolutely continuous with respect to μ then the same is for m as well. But this in view of the Radon-Nikodym theorem gives the existence of the density function $u \in L^1(T, R)$ such that for $A \in \mathcal{L}$

$$m(A) = \int_A u(t)\,\mu(dt).$$

We shall check that

$$u = ess\inf\{u_\lambda\}.$$

From (2.11) we conclude that for every $\lambda \in \Lambda$ we have

$$0 \leq u \leq u_\lambda \leq a,$$

what shows the condition $i.$ of definition.

To see that the condition $ii.$ holds take any $v : T \longrightarrow \overline{R}$ such that for every $\lambda \in \Lambda$ the inequality $u_\lambda \leq v$ is satisfied. Replacing, if necessary, v by $\min\{v, a\}$ we may assume that

$$0 \leq v \leq a.$$

Consider a measure $m_0 : \mathcal{L} \longrightarrow R$ given by

(2.12) $$m_0(A) = \int_A v(t)\,\mu(dt).$$

Then for every $\lambda \in \Lambda$ we have

$$0 \leq m_\lambda \leq m_0$$

and thus

$$0 \leq m \leq m_0.$$

But the latter means that $u \leq v$, what shows $ii.$

The remainder is the condition $iii.$ We recall that μ is separable. Since μ_0 is absolutely continuous with respect to μ, then μ_0 is separable as well. By the Theorem 11 there is a countable $\Lambda_0 \subset \Lambda$, that

$$\inf_{\lambda \in \Lambda}\{m_\lambda\} = m = \inf_{\lambda \in \Lambda_0}\{m_\lambda\}.$$

We shall prove that

$$ess\inf\{u_\lambda\} = u = \inf_{v \in \Lambda_0}\{u_\lambda\}.$$

Of course, since for $\lambda \in \Lambda$ we have $u_\lambda \geq u$, then

$$\inf_{\lambda \in \Lambda_0}\{u_\lambda\} \geq \inf_{\lambda \in \Lambda}\{u_\lambda\} \geq u.$$

Hence for every $A \subset \mathcal{L}$ the following inequalities

$$m\left(A\right) = \int\limits_A \left(\inf_{\lambda \in \Lambda} u_\lambda\left(t\right)\right) \mu\left(dt\right) \mu\left(dt\right) \leq \inf_{\lambda \in \Lambda_0} \int\limits_A \left\{u_\lambda\left(t\right)\right\} \mu\left(dt\right) = m\left(A\right)$$

hold. Hence

$$ess \inf \left\{u_\lambda\right\} = \inf_n \left\{u_{\lambda_n}\right\},$$

what shows *iii*.

Step 2. Let in general $\{u_\lambda\}_{\lambda \in \Lambda} \subset \mathcal{M}\left(T, \overline{R}\right)$ be arbitrary family of measurable functions. For every $k = 1, 2, \ldots$ consider truncated functions $u_{\lambda,k}$ given by

$$(2.13) \qquad u_{\lambda,k}\left(t\right) = \begin{cases} u_\lambda\left(t\right) & if \quad \left|u_\lambda\left(t\right)\right| \leq k \\ k & if \quad u_\lambda\left(t\right) > k \\ -k-1 & if \quad u_\lambda\left(t\right) < -k \end{cases}.$$

Of course, the functions $u_{\lambda,k}$ are measurable with

$$-\left(k+1\right) \leq u_{\lambda,1} \leq \ldots \leq u_{\lambda,k} \leq u_{\lambda,k+1} \leq \ldots \leq \min\left\{u_\lambda, \left(k+1\right)\right\}$$

and

$$(2.14) \qquad \lim_{k \longrightarrow \infty} u_{\lambda,k}\left(t\right) = u_\lambda\left(t\right).$$

By (2.13) the families $\{u_{\lambda,k}\}_{\lambda \in \Lambda}$ are for each $k = 1, 2, \ldots$ integrably bounded by $(k+1)$ and therefore, by the step 1., $u_k = ess \inf \{u_{\lambda,k}\}$ are well defined integrable functions. We claim that $\lim_{k \longrightarrow \infty} u_k$ exists and

$$(2.15) \qquad u = \lim_{k \longrightarrow \infty} u_k = ess \inf \left\{u_\lambda\right\}.$$

By (2.13) and the Theorem 11, for every $k = 1, 2, \ldots$ and any $\lambda \in \Lambda$, we have

$$ess \inf \left\{u_{\lambda,k}\right\} \leq ess \inf \left\{u_{\lambda,k+1}\right\} \leq ess \inf \left\{u_\lambda\right\},$$

or, in other words,

$$u_k \leq u_{k+1} \leq \ldots \leq ess \inf \left\{u_\lambda\right\}.$$

The latter implies that

$$u = \lim_{k \longrightarrow \infty} u_k \leq ess \inf \left\{u_\lambda\right\}.$$

To the end of the proof we need to show that

$$u = \lim_{k \longrightarrow \infty} u_k \geq ess \inf \left\{u_\lambda\right\}.$$

By construction, for every $k = 1, 2, \ldots$ and any $\lambda \in \Lambda$ we have

$$u_{\lambda,k} \geq ess \inf \left\{u_{\lambda,k}\right\} = u_k.$$

Passing to the limit with $k \longrightarrow \infty$ we obtain, by (2.14), that for every $\lambda \in \Lambda$

$$u_\lambda \geq u.$$

Therefore

$$ess\inf\{u_\lambda\} \geq u,$$

what shows (2.15).

The only remaining part to be proved is the condition (iii). For every $k = 1, 2, \ldots$ let $\Lambda_k \subset \Lambda$ be a countable set of indices that

$$(2.16) \qquad ess\inf\{u_{\lambda,k}\} = \inf_{\lambda\in\Lambda_k}\{u_\lambda\}.$$

Take $\Lambda_\infty = \bigcup_{k=1}^{\infty} \Lambda_k$ and observe that it is a denumerable set. We will check that

$$ess\inf\{u_\lambda\} = \inf_{\lambda\in\Lambda_\infty}\{u_\lambda\}.$$

The inequality

$$ess\inf\{u_\lambda\} \leq \inf_{\lambda\in\Lambda_\infty}\{u_\lambda\}$$

is immediate. On the other hand for all $v \in \Lambda$ and $k = 1, 2, \ldots$ we have, by (2.16),

$$(2.17) \qquad u_{\lambda,k} \geq ess\inf\{u_{\lambda,k}\} = \inf_{\lambda\in\Lambda_k}\{u_\lambda\} \geq \inf_{\lambda\in\Lambda_\infty}\{u_\lambda\}.$$

Passing to the limit with $k \longrightarrow \infty$ we obtain for every $\lambda \in \Lambda$, by (2.17), the inequality

$$u_\lambda \geq \inf_{\lambda\in\Lambda_\infty}\{u_\lambda\}$$

and finally

$$ess\inf\{u_\lambda\} \geq \inf_{\lambda\in\Lambda_k}\{u_v\},$$

what completes the proof. $\square$

4. Range of vector measure

4.1. Lapunov theorem. By the range of a vector measure $m \in \mathcal{M}(T, X)$ we call the set

$$\mathcal{R}(m) = \{m(A) : A \in \Sigma\}.$$

A basic result describing the properties of the range belongs to Lapunov [**140**], see cf. [**55**]. Recall that measure $m \in \mathcal{M}(T, X)$ is called nonatomic if every set $A \in \Sigma$ with $|m|(A) > 0$ contains a set $B \in \Sigma$ such that $0 < |m|(B) < |m|(A)$.

THEOREM 13 (Lapunov). *Let* $m = (m_1, ..., m_l) \in \mathcal{M}(T, R^l)$ *be a nonatomic vector measure. Then its range* $\mathcal{R}(m)$ *is convex and compact.*

PROOF. There are many proofs of the theorem. We present one, belonging to Lindenstrauss [150], based on the functional analysis approach.

Consider an integral operator $L : L^1(T, R, |m|) \longrightarrow R$ given by

$$Lh = \int_T f(t) h(t) |m|(dt),$$

where

$$f = \frac{dm}{d|m|} \in L^\infty(T, R, |m|)$$

is the Radon-Nikodym density of m with respect to it's total variation $|m|$. Obviously $L : L^\infty(T, R, |m|) \longrightarrow R$ is a $* - weakly$ continuous functional. Put

$$Z = \{z \in L^\infty(T, R, |m|) : 0 \leq z(t) \leq 1 \text{ a.e. in } T\}$$

and notice that Z is convex and, by the Banach-Alaoglu theorem, $* - weakly$ compact. Now the proof relies on the following relation

$$\mathcal{R}(m) = L(Z).$$

Evidently $m(A) = \int_T \chi_A(t) f(t) |m|(dt) = L\chi_A$ for any $A \in \Sigma$ what shows that $\mathcal{R}(m) \subset L(Z)$. To see the opposite inclusion pick up arbitrarily a point $x \in L(Z)$ and define

$$Z_x = \{z \in Z : L(z) = x\}.$$

It is sufficient to show that Z_x contains some χ_A since then $m(A) = x$. For this purpose notice that Z_x is convex and $* - weakly$ closed, therefore $* - weakly$ compact. Hence by the Krein-Milman theorem Z_x contains an extreme point e. We shall show that e is the characteristic function of a set $A \in \Sigma$. To a contrary assume that there exists $r \in (0, 1)$ such that the set

$$E = \{t : r \leq e(t) \leq 1 - r\}$$

is of positive measure, i.e. $|m|(E) > 0$. Thus the space

$$Y = L^\infty(E, R^l, |m|)$$

is infinitely dimensional and hence $dim(Y) > l$. Therefore there exists a nonzero function $u \in L^\infty(E, R^l, |m|)$ such that $\int_E f(t) u(t) |m|(dt) = 0$ and $|u(t)| < r$ a.e. in E.

Extending u on T by setting $u(t) = 0$ on $T \backslash E$ we obtain a function $u \in L^\infty(T, R^l, |m|)$ such that

$$L(u) = 0$$

and

$$|u(t)| < r \quad a.e. \text{ in } T.$$

One can easily check that $e - u$ and $e + u$ are in Z_x, what contradicts with the extremality of e and completes the proof.   □

There is a long story of results concerning the range of vector measure. The first result of this type belongs to Sierpiński [213] who showed that the range of a real nonatomic measure is a compact interval. The Lapunov Theorem in the form we have presented was shown in the finite dimensional case. In infinitely dimensional spaces the situation is very subtle. The following example shows that there are nonatomic measure with nonconvex and noncompact range.

EXAMPLE 5. *Consider the* $\sigma - field$ $\mathcal{L}_0$ *of the Lebesgue measurable subsets of* $I = [0, 1]$. *Define a measure* $\chi : \mathcal{L}_0 \longrightarrow L^1(I)$ *assigning to each* $A \in \mathcal{L}_0$ *it's characteristic function* χ_A. *Then the range of* χ *is neither convex nor compact.*

PROOF. First of all we shall implement that χ is a measure. For this purpose take an arbitrary decreasing family of measurable sets $\{A_n\}_{n \in N}$ with void intersection. We need to check that

$$\chi_{A_n} \longrightarrow 0 \text{ in } L^1(I).$$

But it is an easy conclusion from the Lebesgue Dominated Convergence Theorem since $\chi_{A_n} \longrightarrow 0$ *a.e. in* T.

The range of χ is nonconvex since $\frac{1}{2}\chi_I \notin \mathcal{R}(\chi) = \{\chi_A : A \in \Sigma\}$.

It can not also be compact. Indeed. Take $B_n = \{r \in I : \sin(2^n \pi r) > 0\}$ for each $n = 1, 2, \dots$. A brief computation shows that for $n \neq m$

$$\left\|\chi_{B_n} - \chi_{B_m}\right\|_1 = \ell(B_n \triangle B_m) = \frac{1}{4}.$$

what is a conclusion of the Proposition 7.   □

Nonatomic measures with convex range in infinitely dimensional spaces are well-characterized. The most powerful result belongs to Knowles (see [133]), but it is beyond of our interest. However we present two very subtle results in infinetely dimensional spaces. Some authors have also consider questions which compact and convex sets can be range of a nonatomic measure (see cf. [131]). About other results concerning vector measures, Lapunov measures, Lapunov extension and related topics we recommend the reader monographs by Knowles and Kluvanek [**133**] and by Diestel and Uhl [**55**] .

The reader might have already noticed that presented above proof of the Lapunov Theorem did take an advantage of the representation (2.2). However a version of this result in infinite dimensional case requires the existence of the density function. Namely, we have:

THEOREM 14 (Hiai & Umegaki, Uhl). *Consider a measure* $m \in M_a(T, X)$ *with the density function* $f \in L^1(T, X, \mu)$. *Then* $cl\mathcal{R}(m)$ *is convex and compact.*

PROOF. We first show that $cl\mathcal{R}(m)$ is compact. Invoking the Theorem 9 we know that operator $L : L^\infty(T, R) \longrightarrow X$ given by

$$Lh = \int_T f(t) h(t) \, d\mu$$

is compact. Therefore the compactness of $cl\mathcal{R}(m)$ follows from inclusion

$$(2.18) \qquad \mathcal{R}(m) = \{m(A) : A \in \Sigma\} = \{L(\chi_A) : A \in \Sigma\} \subset L(Z),$$

where

$$Z = \{z \in L^\infty(T, R, \mu) : 0 \le z(t) \le 1 \quad a.e. \ in \ T\} = B_\infty\left(\frac{1}{2}, \frac{1}{2}\right).$$

For the convexity of $cl\mathcal{R}(m)$ we have to show that for every $a, b \in cl\mathcal{R}(m)$ and $\lambda \in I = [0, 1]$

$$\lambda a + (1 - \lambda) b \in cl\mathcal{R}(m).$$

Fix $a, b \in cl\mathcal{R}(m)$, $\lambda \in I$ and $\varepsilon > 0$. One can choose sets $A, B \in \Sigma$ such that

$$\left| a - \int_A f(t) \mu(dt) \right| < \varepsilon \quad and \quad \left| b - \int_B f(t) \mu(dt) \right| < \varepsilon.$$

Since $f \in L^1(T, X, \mu)$ then there exists a simple function $f_\varepsilon \in L^1(T, X, \mu)$ such that

$$(2.19) \qquad \qquad \|f - f_\varepsilon\|_1 < \varepsilon.$$

Let $f_\varepsilon = \sum\limits_{k=1}^{n} \chi_{A_k} x_k$ for some $x_k \in X$ and a finite partition $\{A_k\}_{k=1}^{n}$ of T. By the construction we have

$$\left| a - \int_A f_\varepsilon(t)\,\mu(dt) \right| < 2\varepsilon \qquad and \qquad \left| b - \int_B f_\varepsilon(t)\,\mu(dt) \right| < 2\varepsilon.$$

Therefore
(2.20)

$$\left| \lambda a + (1-\lambda) b - \left(\lambda \int_A f_\varepsilon(t)\,\mu(dt) + (1-\lambda) \int_B f_\varepsilon(t)\,\mu(dt) \right) \right| < 2\varepsilon$$

or in other words

$$\left| \lambda a + (1-\lambda) b - \left\{ \sum_{k=1}^{n} (\lambda\mu(A \cap A_k) + (1-\lambda)\mu(B \cap A_k)) x_k \right\} \right| < 2\varepsilon$$

Consider measure $\nu : \Sigma \longrightarrow R^n$ given by

$$(2.21) \qquad \nu(C) = (\mu(C \cap A_1), .., \mu(C \cap A_n)).$$

Since ν is nonatomic then the Lapunov Theorem yields the existence of such $C \in \Sigma$ that

$$\lambda\nu(A) + (1-\lambda)\nu(B) = \nu(C).$$

The latter in particular means that for $k = 1, 2, ..., n$ we have

$$\lambda\mu(A \cap A_k) + (1-\lambda)\mu(B \cap A_k) = \mu(C \cap A_k).$$

Therefore

$$\lambda \int_A f_\varepsilon(t)\,\mu(dt) + (1-\lambda) \int_B f_\varepsilon(t)\,\mu(dt) =$$

$$= \sum_{k=1}^{n} x_k (\lambda\mu(A \cap A_k) + (1-\lambda)\mu(B \cap A_k)) =$$

$$= \sum_{k=1}^{n} x_k \mu(C \cap A_k) = \int_C f_\varepsilon(t)\,\mu(dt).$$

By (2.19) we obtain

$$\left| \left(\lambda \int_A f_\varepsilon(t)\,\mu(dt) + (1-\lambda) \int_B f_\varepsilon(t)\,\mu(dt) \right) - \int_C f(t)\,\mu(dt) \right| < \varepsilon,$$

what together with (2.20) gives

$$|\lambda a + (1 - \lambda) b - m(C)| < 3\varepsilon.$$

This in particular means that

$$d(\lambda a + (1 - \lambda) b, \mathcal{R}(m)) < 3\varepsilon.$$

But $\varepsilon > 0$ was arbitrarily chosen, so the latter shows that

$$\lambda a + (1 - \lambda) b \in cl\mathcal{R}(m)$$

and it completes the proof.   □

The previous theorem in a Banach space X with the Radon-Nikodym property can be formulated as follows

COROLLARY 2 (Uhl). *Let X be a Banach space* X *with the Radon-Nikodym property. Then for any given nonatomic measure $m \in \mathcal{M}(T, X)$ the closure of the range $\mathcal{R}(m)$ is compact and convex.*

PROOF. It follows from the representation

$$m(A) = \int_A f(t) |m|(dt),$$

with density function $f \in L^1(T, X, |m|)$.   □

5. Segments for nonatomic measures

5.1. Segments for a measure. In what follows we identify a function $f \in L^1(T, X)$ with the measure $m \in \mathcal{M}_a(T, X)$ given by

$$m(A) = \int_A f(t) \, d\mu.$$

In such case we also write $f = \frac{dm}{d\mu}$. As a conclusion of the Theorem 14 we obtain:

LEMMA 1. *Let $\tilde{m} = (m, m_0) : \Sigma \longrightarrow X \times R^l$ be a nonatomic vector measure with $m \in \mathcal{M}_a$ and $m_0 : \Sigma \longrightarrow R^l$. Then for any given $\varepsilon > 0$ and every $\alpha \in I$ there is a set $A_\alpha \in \Sigma$ such that*

   i. $|m(A_\alpha) - \alpha m(T)| < \varepsilon$
      and
  ii. $m_0(A_\alpha) = \alpha m_0(T).$

PROOF. Notice that $\mathcal{R}(\tilde{m})$ always contains $0 = \tilde{m}(\emptyset)$ and $\tilde{m}(T)$. Thus, by the Theorem 14, the segment

$$[0, \tilde{m}(T)] \subset cl\mathcal{R}(\tilde{m})$$

and hence *i.* holds. A possibility of having an additional condition *ii.* follows from a modification of the proof of the Lapunov Theorem. Namely we should take instead of ν given by (2.21) a measure $\tilde{\nu} = (\nu, m_0) : \Sigma \longrightarrow R^{2n+l}$ and repeat a construction. $\qquad \square$

The previous Lemma 1 gives no information about the family of measurable sets $\{A_\alpha\}_{\alpha \in I}$ satisfying *i.* or *ii.* This can be, however, done in a more precise way. To do that we need some preliminary definitions. Let $(\Lambda, \preceq)$ be an ordered set and consider a family $\{A_\alpha\}_{\alpha \in \Lambda}$ of measurable sets. We say that family the $\{A_\alpha\}_{\alpha \in \Lambda}$ is increasing {decreasing} iff $A_\alpha \subset A_\beta$ $\{A_\alpha \supset A_\beta\}$ whenever $\alpha \preceq \beta$.

DEFINITION 11. *Given a vector measure* $m : \Sigma \longrightarrow X$. *An increasing family* $\{A_\alpha\}_{\alpha \in I}$ *with* $A_0 = \emptyset$, $A_1 = T$ *we call:*
   i. a segment for m *iff for every* $\alpha \in I$ *is*

$$(2.22) \qquad m(A_\alpha) = \alpha m(T);$$

   ii. an ε-*segment for* m *iff for every* $\alpha \in I$ *the inequality*

$$(2.23) \qquad |m(A_\alpha) - \alpha m(T)| < \varepsilon$$

holds.

If $f \in L^1(T, X)$ then by segment or ε-segment for f we mean a corresponding family for measure $m \in \mathcal{M}_a$ with $f = \frac{dm}{d\mu}$.

THEOREM 15. *Let* $\tilde{m} = (m, m_0) : \Sigma \longrightarrow X \times R^l$ *be a nonatomic vector measure with* $m \in \mathcal{M}_a$ *and* $m_0 : \Sigma \longrightarrow R^l$. *Then for every* $\varepsilon > 0$ *there exists a family* $\{A_\alpha\}_{\alpha \in I} \subset \Sigma$ *which is an* ε-*segment for* m *as well as a segment for* m_0.

PROOF. Fix $\varepsilon > 0$. Denote for every $n \in N$

$$\Lambda_n = \left\{ \alpha = \frac{k}{2^n}, k = 1, 2, ..., 2^n \right\}$$

and let $\Lambda = \bigcup_{n=1}^{\infty} \Lambda_n$. First we shall use an induction argument to construct an increasing family $\{A_\alpha\}_{\alpha \in \Lambda_n}$ with $A_0 = \emptyset$, $A_1 = T$ satisfying for every $n = 0, 1, 2, ...$ and all $k = 1, 2, ..., 2^n$ the relations

$$(2.24) \qquad \left| m\left(A_{\frac{k}{2^n}}\right) - \frac{k}{2^n} m(T) \right| < \frac{\varepsilon}{2} \left(1 - \frac{1}{2^n}\right)$$

and

$$(2.25) \qquad m_0\left(A_{\frac{k}{2^n}}\right) = \frac{k}{2^n} m_0(T).$$

For $n = 0$ take $A_0 = \emptyset$ and $A_1 = T$.

For $n = 1$, the Lemma 1 yields the existence of a set $A_{\frac{1}{2}} \in \Sigma$ such that

$$(2.26) \qquad \left| m\left(A_{\frac{1}{2}}\right) - \frac{1}{2}m\left(T\right) \right| < \frac{\varepsilon}{2^2} = \frac{\varepsilon}{2}\left(1 - \frac{1}{2^1}\right)$$

and

$$(2.27) \qquad m_0\left(A_{\frac{1}{2}}\right) = \frac{1}{2}m_0\left(T\right),$$

what means that $\{A_\alpha\}_{\alpha \in \Lambda_1}$ is a required family for $n = 1$.

To make a construction for $n = 2$ let us observe that from (2.26) and (2.27) follows that also for $B_{\frac{1}{2}} = T \backslash A_{\frac{1}{2}}$ the relations

$$\left| m\left(B_{\frac{1}{2}}\right) - \frac{1}{2}m\left(T\right) \right| < \frac{\varepsilon}{2}\left(1 - \frac{1}{2^1}\right)$$

and

$$m_0\left(B_{\frac{1}{2}}\right) = \frac{1}{2}m_0\left(T\right)$$

hold. Consider a measure

$$m_2\left(A\right) = \left(m\left(A \cap A_{\frac{1}{2}}\right), m\left(A \cap B_{\frac{1}{2}}\right), m_0\left(A \cap A_{\frac{1}{2}}\right), m_0\left(A \cap B_{\frac{1}{2}}\right) \right)$$

and observe that

$$m_2 : \Sigma \longrightarrow X \times X \times R^l \times R^l$$

is nonatomic as well. Therefore the Lemma 1 applied to m_2 with $\frac{\varepsilon}{2^3}$ gives the existence of a set $C \in \Sigma$ such that

$$(2.28) \qquad \left| m\left(C \cap A_{\frac{1}{2}}\right) - \frac{1}{2}m\left(A_{\frac{1}{2}}\right) \right| < \frac{\varepsilon}{2^3},$$

$$(2.29) \qquad \left| m\left(C \cap B_{\frac{1}{2}}\right) - \frac{1}{2}m\left(B_{\frac{1}{2}}\right) \right| < \frac{\varepsilon}{2^3},$$

$$(2.30) \qquad m_0\left(C \cap A_{\frac{1}{2}}\right) = \frac{1}{2}m_0\left(A_{\frac{1}{2}}\right)$$

and

$$(2.31) \qquad m_0\left(C \cap B_{\frac{1}{2}}\right) = \frac{1}{2}m_0\left(B_{\frac{1}{2}}\right).$$

Put

$$A_{\frac{1}{4}} = C \cap A_{\frac{1}{2}} \qquad and \qquad A_{\frac{3}{4}} = A_{\frac{1}{2}} \cup \left(C \cap B_{\frac{1}{2}}\right)$$

and observe that family $\{A_\alpha\}_{\alpha \in \Lambda_2}$ is increasing. By (2.28),..., (2.31), one can conclude that

$$\left| m\left(A_{\frac{k}{2^2}}\right) - \frac{k}{2^2}m\left(T\right) \right| < \frac{\varepsilon}{2}\left(1 - \frac{1}{2^2}\right)$$

and

$$m_0\left(A_{\frac{k}{2^2}}\right) = \frac{k}{2^2}m_0\left(T\right),$$

what gives the construction for $n = 2$.

Assume that we have already constructed sets $A_{\frac{k}{2^n}} \in \Sigma$ such that the family $\{A_\alpha\}_{\alpha \in \Lambda_n}$ is increasing and satisfies (2.25) for m_0 and (2.24) for m. Consider measures

$$\widetilde{m}_n = (m_1, ..., m_k, ..., m_{2^n}) \quad and \quad \widetilde{m}_{0n} = (m_{01}, ..., m_{0k}, ..., m_{02^n})$$

with coordinates

$$m_k\left(A\right) = m\left(A \cap B_k\right) \quad and \quad m_{0k}\left(A\right) = m_0\left(A \cap B_k\right),$$

where

$$B_k = A_{\frac{k+1}{2^n}} \backslash A_{\frac{k}{2^n}}.$$

Observe that

$$(\widetilde{m}_n, \widetilde{m}_{0n}) : \Sigma \longrightarrow \underbrace{X \times ... \times X}_{2^n \ times} \times R^{2^n l}$$

is nonatomic as well. Therefore the Lemma 1 applied to $(\widetilde{m}_n, \widetilde{m}_{0n})$ with $\frac{\varepsilon}{2^{n+2}}$ gives the existence of a set $C \in \Sigma$ such that

$$\left|\widetilde{m}_n\left(C\right) - \frac{1}{2}\widetilde{m}_n\left(T\right)\right| < \frac{\varepsilon}{2^{n+2}}$$

and

$$\widetilde{m}_{0n}\left(C\right) = \frac{1}{2}\widetilde{m}_{0n}\left(T\right).$$

The latter means that for $k = 0, 1, ...2^n - 1$ we have

$$(2.32) \qquad \left|m\left(C \cap B_k\right) - \frac{1}{2}m\left(B_k\right)\right| < \frac{\varepsilon}{2^{n+2}}$$

and

$$(2.33) \qquad m_0\left(C \cap B_k\right) = \frac{1}{2}m_0\left(B_k\right).$$

Put

$$A_{\frac{2k+1}{2^{n+1}}} = A_{\frac{k}{2^n}} \cup C \cap B_k, \quad k = 0, 1, ...2^n - 1,$$

and observe that $\{A_\alpha\}_{\alpha \in \Lambda_{n+1}}$ is an increasing family. By (2.32) we conclude that

$$\left|m\left(A_{\frac{k}{2^{n+1}}}\right) - \frac{k}{2^{n+1}}m\left(T\right)\right| < \frac{\varepsilon}{2}\left(1 - \frac{1}{2^n}\right) + \frac{\varepsilon}{2^{n+2}} = \frac{\varepsilon}{2}\left(1 - \frac{1}{2^{n+1}}\right),$$

while from (2.33) we have that

$$m_0\left(A_{\frac{k}{2^{n+1}}}\right) = \frac{k}{2^{n+1}}m_0\left(T\right),$$

what is a construction for $n + 1$.

Having constructed the family $\{A_\alpha\}$ for $\alpha \in \Lambda$ we extend it for every $\alpha \in I$ by

$$A_\alpha = \bigcup_{\frac{k}{2^n} \leq \alpha} A_{\frac{k}{2^n}}.$$

Obviously $\{A_\alpha\}_{\alpha \in \Lambda_{n+1}}$ is increasing. To see that $i.$ and $ii.$ are true take a nondecreasing sequence $\alpha_n = \frac{k_n}{2^n} \longrightarrow \alpha$. Hence

$$A_{\alpha_1} \subset A_{\alpha_2} \subset \dots \subset A_{\alpha_n} \subset \dots,$$

$$A_\alpha = \bigcup_{n=1}^{\infty} A_{\alpha_n}$$

and therefore

$$\widetilde{m}(A_{\alpha_n}) \underset{n \longrightarrow \infty}{\longrightarrow} \widetilde{m}(A_\alpha)$$

But

$$|m(A_{\alpha_n}) - \alpha_n m(T)| < \frac{\varepsilon}{2}\left(1 - \frac{1}{2^n}\right)$$

and thus passing to the limit with $n \longrightarrow \infty$ we get

$$|m(A_\alpha) - \alpha m(T)| \leq \frac{\varepsilon}{2} < \varepsilon.$$

Similarly,

$$m_0(A_\alpha) = \lim_{n \longrightarrow \infty} m_0(A_{\alpha_n}) = \lim_{n \longrightarrow \infty} \alpha_n m_0(T) = \alpha m_0(T),$$

what ends the construction.   □

We shall mention that every finitely dimensional nonatomic vector measure $m : \Sigma \longrightarrow R^l$ admits a segment. Moreover, we may assume that $m_0 = \mu$. In this case the Theorem 14 can also be formutated as follows:

THEOREM 16. *Let $f \in L^1(T, X, \mu)$. Then for every $\varepsilon > 0$ there exists a family $\{A_\alpha\}_{\alpha \in I} \subset \Sigma$ which is an $\varepsilon-$segment for f and a segment for μ.*

Assume now that a measure $m : \Sigma \longrightarrow X$ admits a segment $\{A_\alpha\}_{\alpha \in I} \subset \Sigma$. The smallest $\sigma - field$ containing $\{A_\alpha\}_{\alpha \in I}$ we shall denote by $\Sigma_I \subset \Sigma$. We use the letter I to underline the similarity of the structure Σ_I to the $\sigma - field$ $\mathcal{L}_0$ of the Lebesgue measurable subsets on I. In particular the measure m restricted to Σ_I generates a measure $\ell_X : \mathcal{L}_0 \longrightarrow X$ given by

$$\ell_X((\alpha, \beta]) = m(A_\beta \backslash A_\alpha) = (\beta - \alpha)m(T) = \ell\{(\alpha, \beta]\}m(T)$$

and next extended, by the standard Carathéodory procedure, on $\mathcal{L}_0$. Moreover, a formula

(2.34) $$\tau(\alpha, \beta] = A_\beta \backslash A_\alpha,$$

for $\alpha, \beta \in I$ with $\alpha \leq \beta$, leads to an isomorphism $\tau : (I, \mathcal{L}_0, \ell_X) \longrightarrow (T, \Sigma_I, m)$. So, we have the following

PROPOSITION 17. *Let* $\{A_\alpha\}_{\alpha \in I} \subset \Sigma$ *be a segment for a vector measure* $m : \Sigma \longrightarrow X$. *Then the family* $\{A_\alpha\}_{\alpha \in I}$ *generates a* σ − *field* $\Sigma_I \subset \Sigma$ *and a vector measure* $\ell_X : \mathcal{L}_0 \longrightarrow X$ *such that* $(I, \mathcal{L}_0, \ell_X)$ *and* (T, Σ_I, m) *are isomorphic.*

COROLLARY 3. *Assume that* μ *is a nonnegative, nonatomic measure defined on a* σ − *field* Σ. *Then* Σ *is not compact.*

PROOF. By the Theorem 16 the measure μ admits a segment $\{A_\alpha\}_{\alpha \in I} \subset \Sigma$. Hence, by the Proposition 17, $\{A_\alpha\}_{\alpha \in I} \subset \Sigma$ generates a σ − *field* $\Sigma_I \subset \Sigma$ and an isomorphism

$$\tau : (I, \mathcal{L}_0, \ell) \longrightarrow (T, \Sigma_I, \mu).$$

Take $B_n = \{r \in I : \sin(2^n \pi r) > 0\}$ for each $n = 1, 2, \ldots$ and denote

$$C_n = \tau(B_n).$$

Employing the arguments from the Example 5 we have for $n \neq m$

$$\mu(C_n \triangle C_m) = \ell(B_n \triangle B_m) = \frac{1}{4}.$$

Thus Σ is not compact. $\qquad\square$

COROLLARY 4. *Let* T *is a complete separable metric space with a* σ − *field* $\mathcal{L}$ *of Lebesgue measurable sets given by a finite Radon measure* μ. *Then for any given* $0 \neq u \in L^p(T)$, $p < \infty$, *the set*

$$U = \{u \chi_A : A \in \mathcal{L}\}$$

can not be compact in $L^p(T)$.

PROOF. Denote by $B = \{t : u(t) \neq 0\}$ and observe that

$$U = \{u \chi_{A \cap B} : A \in \mathcal{L}\}.$$

Let us notice that since $\mu(B) > 0$ then the mapping

$$u \chi_{A \cap B} \longrightarrow \chi_{A \cap B}$$

is well defined and continuous from U onto $\mathcal{L}|_B$. Thus if U was compact, therefore $\mathcal{L}_B$ would be compact as well, what, in view of the Corollary 3, gives a contradiction. $\qquad\square$

REMARK 3. *Similarly as in the Proposition 17 we can construct* $\sigma - fields$ $\Sigma_n, \Sigma_N, \Sigma_R \subset \Sigma$ *which are isomorphic with the Lebesgue measurable subsets, respectively, of the cube* $[0,1]^n$, $[0,1]^N$ *or* $[0,1]^R$.

5.2. Segments for a family of measures. Consider a family of measures $\{m_\alpha\}_{\alpha \in \Lambda} \subset \mathcal{M}_a(T, X)$. Then each of them admits an ε-segment. We may also assume in any case that they are also segments for a given nonatomic measure $\mu_0 \in \mathcal{M}(T, R^l)$. In certain situations ε-segments can be however chosen independently on a parameter. Namely:

PROPOSITION 18. *Let S be a compact Hausdorff topological space and consider a continuous mapping $s \longrightarrow m_s$ from S into $\mathcal{M}_a(T, X)$. Then for every $\varepsilon > 0$ there exists a family $\{A_\alpha\}_{\alpha \in I}$ which is an ε-segment for every m_s. If additionally we have given a nonatomic measure $\mu_0 \in \mathcal{M}(T, R^l)$ then we may also require that $\{A_\alpha\}_{\alpha \in I}$ is a segment for μ_0.*

PROOF. Fix $\varepsilon > 0$. For any given $s_0 \in S$ consider a set

$$V_{s_0} = \left\{ s : |m_s - m_{s_0}| < \frac{\varepsilon}{3} \right\}.$$

Observe that family $\{V_{s_0}\}_{s_0 \in S}$ is an open covering of the compact space S. Therefore there exists $s_1, s_2, ..., s_r$ such that

$$S = V_{s_1} \cup ... \cup V_{s_r}.$$

By the *Theorem 14* there exists a family $\{A_\alpha\}_{\alpha \in I}$ which is an $\frac{\varepsilon}{3}$-segment for a measure

$$\widetilde{m} = (m_{s_1}, ..., m_{s_r}, \mu_0) \in \mathcal{M}_a(T, X^r \times R^l)$$

and therefore for every m_{s_i}, $i = 1, 2, ..., r$ and for μ_0. We shall check that $\{A_\alpha\}_{\alpha \in I}$ is a required family. By construction we need only to verify that for every $s \in S$ the following inequality

(2.35) $|m_s(A_\alpha) - \alpha m_s(T)| < \varepsilon$

holds. To see this take any $s \in S$ and let s_i be such that $s \in V_{s_i}$. Since $\{A_\alpha\}_{\alpha \in I}$ which is an $\frac{\varepsilon}{3}$-segment for a measure m_{s_i} then we have

$$|m_{s_i}(A_\alpha) - \alpha m_{s_i}(T)| < \frac{\varepsilon}{3}.$$

Hence

$$|m_s(A_\alpha) - \alpha m_s(T)| \leq$$
$$\leq |m_s(A_\alpha) - m_{s_i}(A_\alpha)| + |m_{s_i}(A_\alpha) - \alpha m_{s_i}(T)| + \alpha |m_s(T) - m_{s_i}(T)| \leq$$
$$\leq 2|m_s - m_{s_i}| + |m_{s_i}(A_\alpha) - \alpha m_{s_i}(T)| < \varepsilon,$$

what shows (2.35) and completes the proof. $\square$

Of course a continuous family $s \longrightarrow m_s$ from S into $\mathcal{M}_a$ may have plenty of ε-segments. In further considerations we need a way to pass continuously from one ε-segment to another. This situation explains

PROPOSITION 19. *Let* $\{A_\alpha\}_{\alpha \in I}$ *and* $\{B_\alpha\}_{\alpha \in I}$ *be two families of measurable sets which are* ε-*segments for* $m \in \mathcal{M}_a$ *and segments for a given nonatomic measure*

$$\tilde{\mu} = (\mu_0, \mu) \in \mathcal{M}\left(T, R \times R^l\right).$$

Then there exists a continuous mapping $D : I \times I \longrightarrow \mathcal{L}$ *with the following properties:*

    i. $D(0, \alpha) = A_\alpha$ *and* $D(1, \alpha) = B_\alpha$ *for every* $\alpha \in I$;

    ii. $\{D(z, \cdot)\}_{z \in I}$ *is for every* $z \in I$ *an* $\varepsilon-$*segment for* m *as well as for* $\tilde{\mu}$;

    iii. $\mu(D(z, \alpha) \triangle D(y, \alpha)) \leq |z - y| \, \mu(T)$.

PROOF. Since $\{A_\alpha\}_{\alpha \in I}$ and $\{B_\alpha\}_{\alpha \in I}$ are in particular segments for μ, therefore the mappings $\alpha \longrightarrow \chi_{A_\alpha}$ and $\alpha \longrightarrow \chi_{B_\alpha}$ as well as $\alpha \longrightarrow m(A_\alpha)$ and $\alpha \longrightarrow m(B_\alpha)$ are continuous. Hence

$$a = \max\left\{ \max_{\alpha \in I} |m(A_\alpha) - \alpha m(T)|, \max_{\alpha \in I} |m(B_\alpha) - \alpha m(T)| \right\} < \varepsilon$$

and we may choose $\eta > 0$ such that

(2.36) $$a + 2\eta < \varepsilon.$$

Consider the measures

$$m_\alpha = \left(m_{|A_\alpha}, m_{|B_\alpha}\right) \in \mathcal{M}_a\left(T, X^2\right)$$

and observe that the mapping $\alpha \longrightarrow m_\alpha$ is continuous. The Proposition 18 yields the existence of a family $\{C_z\}_{z \in I}$ which is an $\eta-$segment for all m_α as well as a segment for $\tilde{\mu}$. Put

$$D(z, \alpha) = \{B_\alpha \cap C_z\} \cup \{A_\alpha \cap (T \backslash C_z)\}.$$

We claim that $D(\cdot, \cdot)$ is a required mapping.

    Obviously *i.* holds.

    To show that *ii.* is satisfied observe that for every $\alpha, z \in I$ we have

(2.37) $$|m_\alpha(C_z) - z m_\alpha(T)| < \eta$$

and thus

$$|m_\alpha(T \backslash C_z) - (1 - z) m_\alpha(T)| < \eta,$$

The condition (2.37) we can rewrite as

$$|m(C_z \cap B_\alpha) - z m(B_\alpha)| < \eta.$$

But $\{B_\alpha\}_{\alpha \in I}$ is an ε-segment for m, therefore

$$|m(C_z \cap B_\alpha) - z\alpha m(T)| \leq$$

$$(2.38) \quad \leq |m(C_z \cap B_\alpha) - zm(B_\alpha)| + |zm(B_\alpha) - z\alpha m(T)| < \eta + za.$$

Similarly from (2.37) we have

$$(2.39) \qquad |m\{(T\backslash C_z) \cap A_\alpha\} - (1-z)\alpha m(T)| < \eta + (1-z)a.$$

Adding (2.38) and (2.39) we get that for all $\alpha \in I$

$$|m\{D(z,\alpha)\} - \alpha m(T)| < 2\eta + a < \varepsilon$$

holds. The calculations for $\tilde{\mu}$ are exactly the same. So $ii.$ holds.

The property $iii.$ follows from the fact that

$$D(z,\alpha) \triangle D(y,\alpha) = (A_\alpha \triangle B_\alpha) \cap (C_z \triangle C_y).$$

It remains to explain the continuity of $D : I \times I \longrightarrow \mathcal{L}$. But this follows from the costruction since the mapping $z \longrightarrow C_z$ is continuous as well as $\alpha \longrightarrow A_\alpha$ and $\alpha \longrightarrow B_\alpha$. This completes the proof. $\square$

REMARK 4. *The mapping $D(\cdot, \cdot)$ appearing in the Proposition 19 plays the role of a homotopy. We shall call D to be a mapping joining $\{A_\alpha\}_{\alpha \in I}$ and $\{B_\alpha\}_{\alpha \in I}$.*

The Proposition 19 gives the existence of ε-segments for a continuous family of functions $s \longrightarrow f_s$ from a compact topological space S into $L^1(T, X)$. In particular the ε-segments exist for a convergent sequence of $f_n \in L^1(T, X)$. If $\{f_n\} \subset L^1(T, X)$ is an arbitrary sequence the latter may fail even if $X = R$. But in this case we can build up a "tower" of ε-segments for each of the functions $g_n = (f_1, ..., f_n) \in L^1(T, X^n)$. Namely we have the following:

THEOREM 17. *Let $\{f_n\}_{n=1}^\infty \subset L^1(T, X)$ be a given sequence and denote by $g_n = (f_1, ..., f_n) \in L^1(T, X^n)$. Then there exists a continuous mapping $D : [0, \infty) \times I \longrightarrow \mathcal{L}$ with the following properties:*

i. $\{D(z,\alpha)\}_{\alpha \in I}$ is for every $z \in [0, \infty)$ an ε-segment for g_n with $n = [z]$;

ii. $\mu(D(z,\alpha) \triangle D(y,\alpha)) \leq |z - y| \mu(T)$.

If, additionally, we have given a nonatomic measure $\mu_0 \in \mathcal{M}(T, R^l)$ then we may require that

iii. $\{D(z,\alpha)\}_{\alpha \in I}$ is for every $z \in [0, \infty)$ an ε-segment for μ_0.

PROOF. Fix $\varepsilon > 0$. By the Theorem 15 for every $n = 0, 1, ...$ there exist families $\{D(n,\alpha)\}_{\alpha \in I}$ which are ε-segments for g_n such that $i.$ holds for $z = n$. We may also require that these families are segments for μ_0. We extend $D(\cdot, \alpha)$ for each $z \in [0, \infty)$. To do so let us first observe that $\{D(m,\alpha)\}_{\alpha\in I}$ are $\varepsilon-$segments for g_n for every $m \geq n+1$. Therefore the Proposition 19 applied for $\{D(n,\alpha)\}_{\alpha\in I}$ and $\{D(n+1,\alpha)\}_{\alpha\in I}$ yields the existence of a continuous mapping C_n : $I \times I \longrightarrow \mathcal{L}$ joining $\{D(n,\alpha)\}_{\alpha\in I}$ and $\{D(n+1,\alpha)\}_{\alpha\in I}$. One can easily check that the mapping $D(z,\alpha)$ defined by $D(z,\alpha) = C_n(z-[z],\alpha)$ for $z \in [n,n+1)$ satisfies $i.$ and $ii.$ If, additionally, all families $\{D(n,\alpha)\}_{\alpha\in I}$ are segments for μ_0 so the same holds for each $\{D(z,\alpha)\}_{\alpha\in I}$ giving $iii.$ $\square$

5.3. Continuous partitions of a measure space. Let $(T,\mathcal{L},\mu)$ be a measure space and (S,d) be a separable metric space. For every $s \in S$ we can consider a partition $\{A_n(s)\}_{n=1}^{\infty}$ of T. Such a family we call finite if it possess this property for every $s \in S$. If each $A_n : S \longrightarrow \mathcal{L}$ is continuous then we say that $\{A_n(s)\}_{n=1}^{\infty}$ is a continuous family of partitions. Similarly, if it is established on S a $\sigma-$field Σ then $\{A_n(s)\}_{n=1}^{\infty}$ is called a $\Sigma-$measurable family of partitions whenever each $A_n : S \longrightarrow \mathcal{L}$ is $\Sigma - measurable.$

THEOREM 18. *Let $p_n : S \longrightarrow L^1(T,X)$ be a sequence of continuous mappings. Consider a locally finite open covering $\{V_n\}_{n=1}^{\infty}$ of S and let $\{\varphi_n\}_{n=1}^{\infty}$ be a partition of unity subordinated to this covering, i.e.*

$$supp\varphi_n \subset V_n, \ for \ n = 1,2,....$$

Then for every continuous $\varepsilon : S \longrightarrow R^+$ there exists a finite and continuous family $\{A_n(s)\}_{n=1}^{\infty}$ of partitions such that for every $s \in S$ the inequality

$$\left| \sum_{n=1}^{\infty} \varphi_n(s) \int_T p_n(s)(t) \mu(dt) - \sum_{n=1}^{\infty} \int_{A_n(s)} p_n(s)(t) \mu(dt) \right| < \varepsilon(s)$$

holds.

If, additionally, we have given a nonatomic measure $\widetilde{\mu} = (\mu_0,\mu) \in \mathcal{M}(T,R^{l+1})$ then we may required that $\{A_n(s)\}_{n=1}^{\infty}$ satisfies, for every $s \in S$, an additional condition

$$|\widetilde{\mu}\{A_n(s)\} - \varphi_n(s)\widetilde{\mu}(T)| < \varepsilon(s).$$

PROOF. Replacing for given continuous $\varepsilon : S \longrightarrow R^+$ the functions p_n with $\frac{1}{\varepsilon}p_n$ and measure $\widetilde{\mu}$ with $\frac{1}{\varepsilon}\widetilde{\mu}$ we may require that

$$\varepsilon(s) \equiv 1.$$

Denote by $N_s = \{s : \varphi_n(s) > 0\}$. Consider functions $h_n : S \longrightarrow I$ such that $h_n(s) \equiv 1$ on $supp \, \varphi_n$ and $supp \, h_n \subset V_n$ for $n = 1,2,....$. Observe that each h_n is continuous and cardinality $card\{N_s\} \leq h_n(s)$. Set

$r\left(s\right) = \sum\limits_{n=1}^{\infty} h_n\left(s\right)$ and $k_n\left(s\right) = r\left(s\right) h_n\left(s\right) p_n\left(s\right)$. Obviously $r\left(s\right) > 0$ and the mappings $r\left(\cdot\right)$ and $k_n\left(\cdot\right)$ are continuous.

Fix $s_0 \in S$ and define

$$(2.40) \quad U_{s_0} = \bigcap_{n \in N_{s_0}} \left\{ s : \begin{array}{c} \varphi_n\left(s\right) > 0, \|k_n\left(s\right) - k_n\left(s_0\right)\|_1 < \frac{1}{16r\left(s_0\right)} \\ \text{and } 3r\left(s\right) < 4r\left(s_0\right) \end{array} \right\}.$$

The family $\{U_s\}_{s \in S}$ is an open covering of the space S. So there exists a sequence of functions $\alpha_m : S \longrightarrow I$ such that the family $\{supp\, \alpha_m\}_{m=1}^{\infty}$ is a locally finite subcovering to $\{U_s\}_{s \in S}$ and the sets

$$W_m = \{s \in S : r_m\left(s\right) = 1\}$$

still cover S. Select s_m such that $W_m \subset U_{s_m}$ and consider functions $u_j \in L^1\left(T, X\right)$ defined by $u_j = k_n\left(s_m\right)$ if $j = 2^m 3^n$ and $u_j = 1$ otherwise. The Theorem 17 applied to sequence $\{u_j\}$ yields the existence of a continuous mapping $D : [0, \infty) \times I \longrightarrow \mathcal{L}$ with the following properties:

1. $\{D\left(z, \alpha\right)\}_{\alpha \in I}$ is for every $z \in [0, \infty)$ an $\frac{1}{2}$-segment for any $u_1, u, ..., u_j$ with $j = [z]$;

2. $\mu_0 \{D\left(z, \alpha\right) \triangle D\left(y, \alpha\right)\} \le |z - y| \mu_0\left(T\right)$.

If, additionally, we have given a nonatomic measure $\mu_0 \in \mathcal{M}\left(T, R^l\right)$ then we may required that

3. $\{D\left(z, \alpha\right)\}_{\alpha \in I}$ is for every $z \in [0, \infty)$ an $\frac{1}{2}$-segment for $\widetilde{\mu} = \left(\mu_0, \mu\right)$.

Let

$$\tau\left(s\right) = \sum\limits_{n=1}^{\infty} r_m\left(s\right) h_n\left(s\right) 2^m 3^n$$

and notice that $\tau\left(\cdot\right)$ is continuous. Set $A\left(s, \alpha\right) = D\left(\tau\left(s\right), \alpha\right)$ and observe that for any $s \in S$ the families $\{A\left(s, \alpha\right)\}_{\alpha \in I}$ are $\frac{1}{4}$-segments for all $u_1, u_2, ..., u_j$ with $j = [\tau\left(s\right)]$. Moreover we may require that they are also $\frac{1}{2}$-segments for $\widetilde{\mu} = \left(\mu_0, \mu\right)$.

We claim that if for some $s \in S$ we have $h_n\left(s\right) = 1$ then $\{A\left(s, \alpha\right)\}_{\alpha \in I}$ is an $\frac{1}{2}$-segment for $r\left(s\right) p_n\left(s\right)$, what in this case means that for every $\alpha \in I$ the following inequality

$$(2.41) \quad \left| \int\limits_{A(s,\alpha)} p_n\left(s\right)\left(t\right) \mu\left(dt\right) - \alpha \int\limits_{T} p_n\left(s\right)\left(t\right) \mu\left(dt\right) \right| < \frac{1}{2r\left(s\right)}$$

holds. Indeed, fix $s \in S$ such that $h_n\left(s\right) = 1$ and $\alpha \in I$. Pick m such that $s \in W_m \subset V_{s_m}$. Hence $[\tau\left(s\right)] \ge 2^m 3^n$ and therefore $A\left(s, \alpha\right)_{\alpha \in I}$ is an $\frac{1}{4}$-segmentt for $k_n(s_m) = r(s_m) p_n(s_m)$. The latter means that

$$\left| \int_{A(s,\alpha)} p_n(s_m)(t)\,\mu(dt) - \alpha \int_T p_n(s_m)(t)\,\mu(dt) \right| < \frac{1}{4r(s_m)}.$$

But

$$(2.42) \qquad \left| \int_{A(s,\alpha)} p_n(s)(t)\,\mu(dt) - \alpha \int_T p_n(s)(t)\,\mu(dt) \right| \le$$

$$\le \left| \int_{A(s,\alpha)} p_n(s_m)(t)\,\mu(dt) - \alpha \int_T p_n(s_m)(t)\,\mu(dt) \right| + 2\,\|k_n(s) - k_n(s_m)\|_1\,.$$

By (2.40) and the choise of s_m the last expression is smaller that

$$\frac{3}{8r(s_m)} < \frac{1}{2r(s)},$$

what shows our claim.

Denote by $z_0(s) = 0$ and $z_n(s) = \varphi_1(s) + \dots + \varphi_n(s)$ and put

$$A_n(s) = A(s, z_n(s)) \setminus A(s, z_{n-1}(s)).$$

Since $\{\varphi_n\}_{n=1}^{\infty}$ is a locally finite and continuous partition of unity then $\{A_n(s)\}_{n=1}^{\infty}$ is a continuous family of partitions of T. Moreover, the mappings $A_n : S \longrightarrow \mathcal{L}$ are continuous because such is the correspondence $(s, \alpha) \longrightarrow A(s, \alpha)$. We shall show that $\{A_n(s)\}_{n=1}^{\infty}$ satisfies 1. To see this observe that from (2.35) and (2.42) follows that for $n = 1, 2, \dots$ we have

$$\left| \int_{A_n(s)} p_n(s)(t)\,\mu(dt) - \varphi_n(s) \int_T p_n(s)(t)\,\mu(dt) \right| < \frac{1}{r(s)}.$$

But the cardinality $card\{s : B_n(s) \ne \emptyset\} = card\{N_s\} \le h_n(s) \le r(s)$ so we obtain

$$\left| \sum_{n=1}^{\infty} \int_{A_n(s)} p_n(s)(t)\,\mu(dt) - \sum_{n=1}^{\infty} \varphi_n(s) \int_T p_n(s)(t)\,\mu(dt) \right| < 1.$$

The additional condition is immediate. This ends the proof. $\qquad \square$

Part 2

MULTIFUNCTIONS

CHAPTER 3

Preliminary notions

The main object examined in this book is a multifuction or, in other words, a multivalued mapping $P : T \longrightarrow N(X)$. We have to report that some authors consider also mappings $P : T \longrightarrow 2^X$ introducing it's domain $T_P = \{t \in T : P(t) \neq \emptyset\}$ but such approach leads to a multifunction $P : T_P \longrightarrow N(X)$.

By the graph of a multifunction $P : T \longrightarrow N(X)$ it is usually meant the set

$$grP = \{(t, x) : x \in P(t)\}.$$

Any pointwise function $p : T \longrightarrow X$ can be identified with a multifunction $P(t) = \{p(t)\}$. On the other hand any pointwise function $p : T \longrightarrow X$ is called a selection of $P : T \longrightarrow N(X)$ iff for every $t \in T$ the relation

$$p(t) \in P(t)$$

holds. If, additionally, p is measurable we shall call it a measurable selection, while for continuous p - a continuous selection.

Many important regularity notions such as measurability and continuity of pointwise mappings have their counterparts for multifunctions. Adequate approach can be braught by a suitable understanding of counter images. For multifunctions we can do that in two ways. Namely, for given $A \subset X$ we can define

$$P^-(A) = \{t : P(t) \cap A \neq \emptyset\} \qquad and \qquad P^+(A) = \{t : P(t) \subset A\}.$$

From the definition one can easily check that these objects satisfy the following formulas:

$$P^-(X \backslash A) = T \backslash P^+(A), \qquad P^+(X \backslash A) = T \backslash P^-(A),$$

(3.1)

$$P^-\left(\bigcup_{\alpha \in \Lambda} A_\alpha\right) = \bigcup_{\alpha \in \Lambda} P^-(A_\alpha), \qquad P^-\left(\bigcap_{\alpha \in \Lambda} A_\alpha\right) \subset \bigcap_{\alpha \in \Lambda} P^-(A_\alpha),$$

(3.2)

$$\bigcup_{\alpha \in \Lambda} P^+(A_\alpha) \subset P^+\left(\bigcup_{\alpha \in \Lambda} A_\alpha\right), \qquad P^+\left(\bigcap_{\alpha \in \Lambda} A_\alpha\right) = \bigcap_{\alpha \in \Lambda} P^+(A_\alpha),$$

where Λ is an index set.

If $A \subset B$ then

$$P^- (A) \subset P^- (B) \qquad and \qquad P^+ (B) \subset P^+ (A).$$

The reader may also easily verify that:

(i) for every closed $F \subset X$

$$P^- (F) = \{t : P (t) \subset F\} = \{t : clP (t) \subset F\},$$

(ii) for every open $U \subset X$

$$(3.3) \qquad P^- (U) = \{t : P (t) \cap U \neq \emptyset\} = \{t : clP (t) \cap U \neq \emptyset\}$$

and

(iii) for $P (t) = \{p (t)\}$, where $p : T \longrightarrow X$, we have

$$P^- (A) = P^+ (F) = p^{-1} (A).$$

For a multifunction $P : T \longrightarrow N (X)$ by the image of a set $A \subset T$ we mean a set

$$P (A) = \bigcup_{x \in A} P (x).$$

For a given $\omega : X \longrightarrow Y$ by $\omega (P) = \omega \circ P$ we denote the compose multifunction given by

$$\omega (P) (t) = (\omega \circ P) (t) = \omega (P (t)).$$

Notice that

$$(3.4) \qquad (\omega (P))^- (A) = P^- (\omega^{-1} (A)),$$

$$(\omega (P))^+ (A) = P^+ (\omega^{-1} (A))$$

and

$$(\omega (P)) (A) = \omega (P (A)).$$

CHAPTER 4

Upper and lower semicontinuous multifunctions

1. General properties

Let T, X and Y be topological spaces. We begin with the following:

DEFINITION 12. *A multifunction $P : T \longrightarrow N(X)$ is said to be*
  (1) *upper semicontinuous (u.s.c.) at t_0 iff t_0 is an interior point in $P^+(V)$ for every open V such that $t_0 \in P^+(V)$;*
  (2) *lower semicontinuous (l.s.c.) at t_0 iff t_0 is an interior point in $P^-(V)$ for every open V such that $t_0 \in P^-(V)$;*
  (3) *continuous at t_0 if it is both l.s.c. and u.s.c. at t_0.*

Lower and upper semicontinuity can also be characterized in terms of generalized sequences. Namely we have the following:

PROPOSITION 20. *Consider a multivalued mapping $P : T \longrightarrow N(X)$. Then*
  (1) *P is l.s.c. at t_0 iff for every $x_0 \in P(t_0)$ and each generalized sequence $t_\alpha \longrightarrow t_0$ one can pick up $x_\alpha \in P(t_\alpha)$ such that $x_\alpha \longrightarrow x$;*
  (2) *if P is u.s.c. at t_0 then for every net $(t_\alpha, x_\alpha) \in grP$ such that $(t_\alpha, x_\alpha) \longrightarrow (t_0, x_0)$ one has $(t_0, x_0) \in grP$. If T is compact then both conditions are equivalent.*

PROOF. (1) $\Longrightarrow$
Take an arbitrary $x_0 \in P(t_0)$ and any $t_\alpha \longrightarrow t_0$. Choose a partially ordered family $\{V_\beta\}_{\beta \in \Lambda}$ of open neighbourhoods of x_0 such that $V_{\tilde{\beta}} \subset V_\beta$ for $\tilde{\beta} \succeq \beta$ and $\bigcap_{\beta \in \Lambda} V_\beta = \{x_0\}$. By the l.s.c. at t_0 there is an interior point in $P^-(V_\beta)$. Hence there are open U_β such that $t_0 \in U_\beta \subset P^-(V_\beta)$. Thus for every $t \in U_\beta$ we have $P(t) \cap V_\beta \neq \emptyset$. But $t_\alpha \longrightarrow t_0$, so for every $\beta \in \Lambda$ there is $\alpha(\beta)$ with the property that for $\alpha \succeq \alpha(\beta)$ we have $t_\alpha \in U_\beta$. We may assume that $\alpha(\beta)$ is an increasing function. Hence for $\alpha \succeq \alpha(\beta)$ there is $P(t_\alpha) \cap V_\beta \neq \emptyset$. Choose arbitrary $x_\alpha \in P(t_\alpha) \cap V_\beta$ for $\alpha \succeq \alpha(\beta)$ and $\alpha \underset{\sim}{\succeq} \alpha(\tilde{\beta})$ for $\tilde{\beta} \succeq \beta$. One can easily observe that $x_\alpha \longrightarrow x_0$ is the required sequence.

59

(1) $\Longleftarrow$

To a contrary assume that P is not *l.s.c.* at t_0. Therefore there is an open V with $P(t_0) \cap V \neq \emptyset$ such that t_0 is not an interior point in $P^-(V)$. Thus we may find a generalized sequence

$$\{t_\alpha\}_{\alpha \in \Lambda} \subset T \backslash P^-(V) = P^+(X \backslash V)$$

such that $t_\alpha \longrightarrow t_0$. Take $x_0 \in P(t_0) \cap V$ and select for every $\alpha \in \Lambda$ such $x_\alpha \in P(t_\alpha) \subset (X \backslash V)$ that $x_\alpha \longrightarrow x_0$. Then $x_0 \in X \backslash V$, what is a contradiction with the choice of x_0.

(2) $\Longrightarrow$

Let $\{(t_\alpha, x_\alpha)\}_{\alpha \in \Lambda} \subset grP$ be an arbitrary net converging to (t_0, x_0). We need to show that $(t_0, x_0) \in grP$. To a contrary assume that $(t_0, x_0) \notin grP$. Thus $x_0 \notin P(t_0)$. Since $P(t_0)$ is closed, then there exists an open $V \ni x_0$ such that $P(t_0) \cap \overline{V} = \emptyset$ or, equivalently, $t_0 \in P^+(X \backslash \overline{V})$. But $P^+(X \backslash \overline{V})$ is open. Therefore there exists $\alpha_0 \in \Lambda$ such that for $\alpha \succeq \alpha_0$ we have $t_\alpha \in P^+(X \backslash \overline{V})$ and $x_\alpha \in V$. Hence $x_\alpha \in P(t_\alpha) \subset X \backslash \overline{V}$, a contradiction.

(2) $\Longleftarrow$

To a contrary assume that P is not *u.s.c.* at t_0. Therefore there is an open $V \supset P(t_0)$ such that t_0 is not an interior point in $P^+(V)$. Thus we may find $\{t_\alpha\}_{\alpha \in \Lambda} \subset T \backslash P^+(V) = P^-(X \backslash V)$ and $t_\alpha \longrightarrow t_0$. Choose $\{x_\alpha\}_{\alpha \in \Lambda} \subset P(t_\alpha) \cap (X \backslash V)$. By compactness we may extract a converging subnet, say $x_{\tilde{\alpha}} \longrightarrow x_0$, and thus

(4.1) $$x_0 \in X \backslash V.$$

Therefore $(t_\alpha, x_\alpha) \in grP$ and $(t_\alpha, x_\alpha) \longrightarrow (t_0, x_0)$. Hence $(t_0, x_0) \in grP$, i.e. $x_0 \in P(t_0) \subset V$, a contradiction with (4.1). $\square$

REMARK 5. *If T is a separable metric space then in the above proposition we can replace generalized sequences by sequences.*

DEFINITION 13. *A multifunction $P : T \longrightarrow N(X)$ is u.s.c. (l.s.c.) in $D \subset T$ iff it is u.s.c. (l.s.c.) at any $t \in D$. If $D = T$ then we say that P is u.s.c. (l.s.c.).*

Before we present some examples of *l.s.c.* and *l.s.c.* mappings we shall give some useful characterizations of upper and lower semicontinuity.

PROPOSITION 21. *For a multifunction $P : T \longrightarrow N(X)$ the following conditions are equivalent:*

    i. P is u.s.c.;

    ii. for every open $V \subset X$ the set $P^+(V)$ is open in T;

    iii.. for every closed $F \subset X$ the set $P^-(F)$ is closed in T;

iv. each $D \subset X$ an inclusion $cl\,(P^-(D)) \subset P^-(clD)$ is satisfied;

v. for each $D \subset X$ an inclusion $P^+(IntD) \subset IntP^+(D)$ holds.

PROOF. $i. \Longrightarrow ii.$

Fix open $V \subset X$ and take $t_0 \in P^+(V)$. So t_0 is an interior point. But then $P^+(V)$ is open;

$ii. \Longrightarrow iii.$

If $F \subset X$ is closed then $P^-(F) = T \backslash P^+(X \backslash F)$ is closed as well;

$iii. \Longrightarrow iv.$

Since $P^-(clD)$ is closed and $P^-(D) \subset P^-(clD)$ then $iv.$ is true;

$iv. \Longrightarrow v.$

It can be concluded from a sequel

$$P^+(IntD) = P^+(X \backslash cl\,(X \backslash D)) = T \backslash P^-(cl\,(X \backslash D)) \subset$$

$$\subset T \backslash cl\,(P^-(X \backslash D)) = T \backslash cl\,(T \backslash P^+(D)) = IntP^+(D).$$

$v. \Longrightarrow i.$

Since for any open V we have

$$P^+(V) = P^+(IntV) \subset IntP^+(D),$$

then each $t_0 \in P^+(V)$ is an interior point. □

Similarly we can prove the following characterization:

PROPOSITION 22. *For a multifunction* $P : T \longrightarrow N(X)$ *the following conditions are equivalent:*

  i. *P is l.s.c.;*

 ii. *for every open $V \subset X$ the set $P^-(V)$ is open in T;*

iii. *for every closed $F \subset X$ the set $P^+(F)$ is closed in T;*

 iv. *for each $D \subset X$ an inclusion $cl\,(P^+(D)) \subset P^+(clD)$ holds;*

  v. *for each $D \subset X$ an inclusion $P^-(IntD) \subset IntP^-(D)$ holds.*

REMARK 6. *We should also notice that*

(1) $P : T \longrightarrow N(X)$ *is l.s.c. iff* $clP : T \longrightarrow cl\,(X)$ *is l.s.c.. It follows from the fact that for every open U we have*

$$P^-(U) = (clP)^-(U).$$

(2) *A verification in metric spaces (X, d) that a given multifunction $P : T \longrightarrow N(X)$ is l.s.c. can be reduced to a checking that $P^-(B(x, r))$ is open for every open ball $B(x, r)$. It follows from the fact that every open set in X is union of open balls and formula (3.1).*

PROPOSITION 23. *Let $P : T \longrightarrow N(X)$ be a multifunction. Then P is l.s.c. at t_0 iff for any $x_0 \in P(t_0)$ and every $t_\alpha \longrightarrow t_0$ there exist*

$x_\alpha \in P(t_\alpha)$ such that $x_\alpha \longrightarrow x_0$. If T is separable then the same holds for countable sequences.

PROOF. Fix $x_0 \in P(t_0)$ and $t_\alpha \longrightarrow t_0$. Choose a decreasing family $\{U_n\}_{n=1}^\infty$ of open sets such that $\bigcap_{n=1}^\infty U_n = \{x_0\}$. Then the sets $V_n = P^- U_n$ form a decreasing family of open neighbourhoods of t_0. For every $n \in N$ there exists α_n such that for $\alpha \succeq \alpha_n$ we have $t_\alpha \in V_n$ i.e. $P(t_\alpha) \cap V_n \neq \emptyset$. We may assume that $\alpha_1 \preceq \alpha_2 \preceq ... \preceq \alpha_n \preceq ...$. Select elements $x_\alpha \in P(t_\alpha) \cap V_n$ for $\alpha_n \preceq \alpha \prec \alpha_{n+1}$. One can easily check that $x_\alpha \longrightarrow x_0$, what shows our claim.   □

PROPOSITION 24. Let $P : T \longrightarrow cl(X)$ be an u.s.c. multifunction. Then its graph $grP = \{(t,x) : x \in P(t)\}$ is closed in $T \times X$.

PROOF. Let $\{(t_\alpha, x_\alpha)\}_{\alpha \in \Lambda} \subset grP$ be an arbitrary generalized sequence converging to (t,x). We need to show that $(t,x) \in grP$. To a contrary assume that $(t,x) \notin grP$. Thus $x \notin P(t)$. Since $P(t)$ is closed, then there exists an open $V \ni x$ such that $P(t) \cap \overline{V} = \emptyset$ or, equivalently, $t \in P^+(X \backslash \overline{V})$. But $P^+(X \backslash \overline{V})$ and V are open. Therefore, there exists $\alpha_0 \in \Lambda$ such that for $\alpha \succeq \alpha_0$ we have $x_\alpha \in V$ and $t_\alpha \in P^+(X \backslash \overline{V})$. Thus for $\alpha \succeq \alpha_0$ at the same time is $x_\alpha \in P(t_\alpha) \subset X \backslash \overline{V}$ and $x_\alpha \in V$, a contradiction.   □

Having characterized u.s.c. and l.s.c. we can present some their illustrations.

PROPOSITION 25. Given a family $\{p_\alpha\}_{\alpha \in \Lambda}$ of continuous functions from T into a Banach space X. Then a multifunction $P : T \longrightarrow cl(X)$ given by

$$P(t) = cl\{p_\alpha(t) : \alpha \in \Lambda\}$$

is l.s.c. .

PROOF. The conclusion follows from an observation that for arbitrary closed set $F \subset X$ the counter image $P^+(F) = \bigcap_\alpha p_\alpha^{-1}(F)$ is closed.   □

EXAMPLE 6.   (1) A multifunction $P_1 : T \longrightarrow N(X)$ given by $P_1(t) = \{p(t)\}$ is l.s.c. and u.s.c. iff p is continuous;
    (2) A multifunction $P_2 : T \longrightarrow clco(R)$ given by

$$P_2(t) = \begin{cases} \{1\} & for \quad t < 0 \\ \{-1\} & for \quad t > 0 \\ [-1,1] & for \quad t = 0 \end{cases}$$

is u.s.c. but not l.s.c.;

(3) *A multifunction $P_3 : R \longrightarrow cl\,(R)$ given by*

$$P_3\,(t) = \begin{cases} [-1,1] & \text{for} \quad t \neq 0 \\ \{0\} & \text{for} \quad t = 0 \end{cases}$$

   is l.s.c. but not u.s.c.;
(4) *Assume that $Z \subset Y$ is a compact set and $f : T \times Y \longrightarrow X$ is continuous. Then multifunction $P_4 : T \longrightarrow cl\,(X)$ given by $P_4\,(t) = f\,(t,Z)$ is u.s.c. and may not be l.s.c.;*
(5) *Let $P_5\,(t) = [a\,(t),b\,(t)]$, where $a,b : T \longrightarrow R$ are given. Then*
   (a) P_5 *is u.s.c. iff a is u.s.c. and b is l.s.c.;*
   (b) P_5 *is l.s.c. iff a is l.s.c. and b is u.s.c..*
(6) *Let $P_6\,(t) = \overline{B}\,(0,r\,(t)) \subset R^l$, where $r : T \longrightarrow R$ is given. Then*
   (a) P_6 *is u.s.c. iff r is u.s.c.;*
   (b) P_6 *is l.s.c. iff r is l.s.c..*
(7) *Let X be a Banach space and $P_7 : T \longrightarrow N\,(X)$ be l.s.c. Then $coP_7 : T \longrightarrow N\,(X)$ given by the formula $(coP_7)\,(t) = co\,(P_7\,(t))$ is l.s.c. as well.*

PROOF. Examples 1.-4. are straightforward and we leave them as exercises.

Example 5. $\Longrightarrow$.
Observe that for each r we have

$$a^{-1}\,\{(r,\infty)\} = \{t : a\,(t) > r\} =$$
$$= \{t : P_5\,(t) \subset (r,\infty)\} = P_5^+\,\{(r,\infty)\}$$

and

$$b^{-1}\,\{(-\infty,r)\} = \{t : b\,(t) < r\} =$$
$$\{t : P_5\,(t) \subset (-\infty,r)\} = P_5^+\,\{(-\infty,r)\} .$$

If P is u.s.c. then $a^{-1}\,\{(r,\infty)\}$ and $b^{-1}\,\{(-\infty,r)\}$ are open.

$\Longleftarrow$.

Let $V \subset R$ be open. We need to check that any $t_0 \in P_5^+\,(V)$ is an interior point. Assume that $P_5\,(t_0) = [a\,(t_0),b\,(t_0)] \subset V$ and take such a component $(r,s) \subset V$ that $[a\,(t_0),b\,(t_0)] \subset (r,s)$ or, equivalently, $t_0 \in P_5^+\,\{(r,s)\}$. But

$$U = P_5^+\,\{(r,s)\} = a^{-1}\,\{(r,\infty)\} \cap b^{-1}\,\{(-\infty,s)\}$$

is open and $t_0 \in U \subset P_5^+\,(V)$, what ends the proof;

Example 5b. $\Longrightarrow$.
Observe that for each r we have

$$a^{-1}\,\{(-\infty,r)\} = \{t : a\,(t) < r\} =$$
$$= \{t : P_5\,(t) \cap (-\infty,r) \neq \emptyset\} = P_5^-\,\{(-\infty,r)\}$$

and
$$b^{-1}\{(r,\infty)\} = \{t : b(t) > r\} =$$
$$= \{t : P_5(t) \cap (r,\infty) \neq \emptyset\} = P_5^-\{(r,\infty)\}.$$
Therefore, if P is $l.s.c.$ then $a^{-1}\{(r,\infty)\}$ and $b^{-1}\{(-\infty,r)\}$ are open.

$\Longleftarrow$.

Let $V \subset R$ be open. We need to check that $P_5^-(V)$ is open. But

$$V = \bigcup_{n=1}^{\infty} (r_n, s_n).$$

So

$$P_5^-(V) = P_5^-\left(\bigcup_{n=1}^{\infty} (r_n, s_n)\right) = \bigcup_{n=1}^{\infty} P_5^-\{(r_n, s_n)\} =$$

$$= \bigcup_{n=1}^{\infty} a^{-1}\{(-\infty, s_n)\} \cap b^{-1}\{(r_n, \infty)\}$$

is open, what ends the proof.

Example 6a.

$\Longrightarrow$. It is a consequence from the sequel

$$r^{-1}\{(-\infty, h)\} = \{t : r(t) < h\} =$$
$$= \{t : P_6(t) \subset B(0, h)\} = P_6^+\{B(0, h)\}.$$

$\Longleftarrow$. If for some open V and given t_0 we have

$$P_6(t_0) = \overline{B}(0, r(t_0)) \subset V,$$

then, by compactness of the sphere, there is $h > 0$ such that

$$\overline{B}(0, r(t_0)) \subset B(0, h) \subset V.$$

Therefore

$$U = P_6^+\{B(0, h)\} = \{t : r(t) < h\} = r^{-1}\{(-\infty, h)\}$$

is open and

$$t_0 \in U = P_6^+\{B(0, h)\} \subset P_6^+(V).$$

Example 6b.

$\Longrightarrow$. It follows from the formulas

$$r^{-1}\{(-\infty, h]\} = \{t : r(t) \leq h\} =$$
$$= \{t : P_6(t) \subset \overline{B}(0, h)\} = P_6^+\{\overline{B}(0, h)\}.$$

$\Longleftarrow$. Observe that

$$P_6(t) \cap B(x, a) \neq \emptyset \Longleftrightarrow r(t) > |x| - a.$$

So the set
$$P_6^- \{B(x,r)\} = r^{-1} \{(|x| - a, \infty)\}$$
is open. Take an arbitrary open set V and represent it as
$$V = \bigcup_{n=1}^{\infty} B(x_n, a_n).$$
Then the set $P_6^-(V)$ is open, because
$$P_6^-(V) = \bigcup_{n=1}^{\infty} P_6^-(B(x_n, a_n)) = \bigcup_{n=1}^{\infty} r^{-1} \{(|x_n| - a_n, \infty)\}.$$

Example 7. We shall apply the Proposition 23. Fix t_0, $t_\alpha \longrightarrow t_0$ and $x_0 \in (coP)(t_0)$. One can find $x_1, x_2, ..., x_n \in P(t_0)$ and $\lambda_1, \lambda_2, ..., \lambda_n \in [0,1]$ with $\sum_{k=1}^{n} \lambda_k = 1$ such that $x_0 = \sum_{k=1}^{n} \lambda_k x_k$. Choose $x_{\alpha,k} \in P(t_\alpha)$ such that $x_{\alpha,k} \longrightarrow x_0$. Then $x_\alpha = \sum_{k=1}^{n} \lambda_k x_{\alpha,k} \in co(P(t_\alpha))$ and $x_\alpha \longrightarrow x_0$, what completes the proof. □

Upper and lower semicontinuous multivalued mappings possess many nice properties and some of them are presented below.

PROPOSITION 26. *Consider l.s.c. multifunctions $P_n : T \longrightarrow N(X)$ and continuous $p_n : T \longrightarrow X$, $n = 0, 1, 2, ...$ and assume that $\omega : X \longrightarrow Y$ and $r : T \longrightarrow (0, \infty)$ are continuous. Then formulas*
$$P(t) = r(t) P_0(t) + p_0(t),$$
$$P(t) = r(t) P_0(t) + p_0(t),$$
$$R(t) = cl\{p_n(t) : n = 0, 1, 2, ...\}$$
and
$$\omega(P)(t) = \omega\{P(t)\}$$
define new l.s.c. multivalued mappings.
Moreover, if $r : T \longrightarrow (0, \infty)$ is such u.s.c. function that
(4.2)
$$S(t) = P_0(t) \cap B(p_0(t), r(t)) \neq \emptyset$$
then S is l.s.c. as well.

PROOF. *a)* Fix $x_0 \in P(t_0)$ and $t_\alpha \longrightarrow t_0$. Then
$$z_0 = \frac{1}{r(t_0)} (x_0 - p_0(t_0)) \in P_0(t_0).$$
By the *l.s.c.* of P_0 there exist $z_\alpha \in P_0(t_\alpha)$ such that $z_\alpha \longrightarrow z_0$. One can easily check that $x_\alpha = r(t_\alpha) z_\alpha + p_0(t_\alpha) \in P(t_\alpha)$ and $x_\alpha \longrightarrow x_0$;

b) *l.s.c.* of Q follows easily from an observation that

$$Q^-(U) = \left\{ t : \left(\bigcup_{n=0}^{\infty} P_n(t) \right) \cap U \neq \emptyset \right\} =$$

$$= \bigcup_{n=0}^{\infty} \{t : P_n(t) \cap U \neq \emptyset\} = \bigcup_{n=0}^{\infty} P_n^-(U);$$

c) *l.s.c.* of R can be concluded from an observation that

$$R(t) = cl \left\{ \bigcup_{n=0}^{\infty} S_n(t) \right\},$$

where $R_n(t) = \{p_n(t)\}$.

d) *l.s.c.* of $\omega(P)$ follows from the fact that $(\omega(P))^-(U) = P^-(\omega^{-1}(U))$;

e) finally to see the *l.s.c.* of S notice that by a) it is enough to assume that $p_0(t) = 0$. Fix $x_0 \in S(t_0)$ and $t_\alpha \longrightarrow t_0$. Then $x_0 \in P_0(t_0)$ and $|x_0| < r(t_0)$. Pick a such that $|x_0| < a < r(t_0)$ and $x_\alpha \in P_0(t_\alpha)$ such that $x_\alpha \longrightarrow x_0$. There exists α_0 such that $|x_\alpha| < a < r(t_\alpha)$ for $\alpha \succeq \alpha_0$. Therefore for $\alpha \succeq \alpha_0$ we have $x_\alpha \in S(t_\alpha)$. □

EXAMPLE 7. *We have shown that S given by (39) is l.s.c.. This however may not be true if we replace open ball by the closed one. For $t \in [0,1]$ denote by $P(t)$ the circle centered at $(t,0)$ and radius $r = t$. Obviously $t \longrightarrow P(t)$ is continuous and*

$$P(t) \cap \overline{B}[(0,0);1] = \left\{ \begin{array}{ll} \{(0,0)\} & if \quad t \in [0,1) \\ \{(0,0),(2,0)\} & if \quad t = 1 \end{array} \right. .$$

Therefore the mapping $t \longrightarrow P(t) \cap \overline{B}((0,0);1)$ is not l.s.c. (but u.s.c.).

2. Upper and lower semicontinuity in metrizable spaces

If (X,d) is a metrizable space then the *l.s.c.* and *u.s.c.* can also be expressed in metric terms. Recall that for any $A, B \in N(X)$ we denote by

$$d_0(A,B) = \sup_{a \in A} d(a,B)$$

and by $d_H(A,B)$ the Hausdorff distance

$$d_H(A,B) = \max \{d_0(A,B), d_0(B,A)\}.$$

PROPOSITION 27. *Consider a multivalued mapping $P : T \longrightarrow cl(X)$. Then*

(1) *P is l.s.c. at t_0 iff $\lim_{t \longrightarrow t_0} d_0(P(t_0), P(t)) = 0$.*

(2)   If P is u.s.c. at t_0 then $\lim\limits_{t \longrightarrow t_0} d_0 \left(P\left(t\right), P\left(t_0\right)\right) = 0$. For a compact (X, d) both conditions are equivalent.

(3) If $P : T \longrightarrow b\left(X\right)$ is continuous at t_0 then $\lim\limits_{t \longrightarrow t_0} d_H \left(P\left(t\right), P\left(t_0\right)\right) = 0$. For a compact (X, d) both conditions are equivalent.

PROOF. (1) $\Longrightarrow$.

Take arbitrary $x_0 \in P\left(t_0\right)$ and any $\varepsilon > 0$. Consider the set $V = P^- \{B\left(x_0, \varepsilon\right)\}$ and observe that V is open and $t_0 \in V$. By l.s.c. at t_0 for $t \in V$ we have

$$d\left(x_0, P\left(t\right)\right) \le d\left(x_0, P\left(t\right) \cap B\left(x_0, \varepsilon\right)\right) < \varepsilon.$$

Hence

$$d_0\left(P\left(t_0\right), P\left(t\right)\right) = \sup_{x_0 \in P(t_0)} d\left(x_0, P\left(t_0\right)\right) \le \varepsilon$$

and this shows that

$$\lim_{t \longrightarrow t_0} d_0\left(P\left(t_0\right), P\left(t\right)\right) = 0.$$

$\Longleftarrow$.

To a contrary assume that $\lim\limits_{t \longrightarrow t_0} d_0\left(P\left(t_0\right), P\left(t\right)\right) = 0$ but P is not l.s.c. at t_0. Therefore there is an open V such that $P\left(t_0\right) \cap V \ne \emptyset$ and t_0 is not an interior point in $P^-\left(V\right)$. Thus we may find a generalized sequence

$$\{t_\alpha\}_{\alpha \in \Lambda} \subset T \backslash P^-\left(V\right) = P^+\left(X \backslash V\right)$$

such that $t_\alpha \longrightarrow l_0$. Take $x_0 \in P\left(t_0\right) \cap V$ and choose for every $\alpha \in \Lambda$ such $x_\alpha \in P\left(t_\alpha\right) \subset (X \backslash V)$ that $d\{x_0, P\left(t_\alpha\right)\} = d\left(x_0, x_\alpha\right)$. Then

$$d\left(x_0, x_\alpha\right) = d\left(x_0, P\left(t_\alpha\right)\right) \le d_0\left(P\left(t_0\right), P\left(t_\alpha\right)\right),$$

what in turn means that $x_\alpha \longrightarrow x_0$. But then $x_0 \in X \backslash V$, what is a contradiction with the choice of x_0.

(2) $\Longrightarrow$.

Take any $\varepsilon > 0$ and let $V = \{x : d\left(x, P\left(t_0\right)\right) < \varepsilon\}$. Then $P^+\left(V\right)$ is open and $t_0 \in P^+\left(V\right)$. Hence there is an open U such that $t_0 \in U \subset P^+\left(V\right)$. Thus for every $t \in U$ we have $P\left(t\right) \subset V$ and therefore

$$d_0\left(P\left(t\right), P\left(t_0\right)\right) = \sup_{x \in P(t)} d\left(x, P\left(t_0\right)\right) \le \sup_{x \in V} d\left(x, P\left(t_0\right)\right) \le \varepsilon.$$

The latter shows that $\lim\limits_{t \longrightarrow t_0} d_0\left(P\left(t\right), P\left(t_0\right)\right) = 0$.

$\Longleftarrow$.

To a contrary assume that $\lim\limits_{t \longrightarrow t_0} d_0\left(P\left(t\right), P\left(t_0\right)\right) = 0$ but P is not u.s.c. a t_0. Therefore there is an open $V \supset P\left(t_0\right)$ such that t_0 is not an interior point in $P^+\left(V\right)$. Thus we may find $\{t_\alpha\}_{\alpha \in \Lambda} \subset T \backslash P^+\left(V\right) =$

$P^{-}(X \backslash V)$ with $t_\alpha \longrightarrow t_0$. Choose $\{x_\alpha\}_{\alpha \in \Lambda} \subset P(t_\alpha) \cap (X \backslash V)$. By compactness we may assume that $x_\alpha \longrightarrow x_0$ and then

$$(4.3) \qquad\qquad x_0 \in X \backslash V.$$

Observe that

$$d(x_\alpha, P(t_0)) \leq d_0(P(t_\alpha), P(t_0)).$$

The latter implies that

$$d(x_0, P(t_0)) = \lim d(x_\alpha, P(t_0)) \leq \lim d_0(P(t), P(t_0)) = 0$$

and hence $x_0 \in P(t_0) \subset V$, a contradiction with (4.3).

(3) It is an easy consequence of (1) and 2. The notions of $l.s.c.$ and $u.s.c.$ can also be characterized in terms of sequences or generalized sequences. $\qquad\qquad\qquad\square$

Let (X, d) be a metric space.

PROPOSITION 28. *For a multifunction $P : T \longrightarrow N(X)$ the following conditions are equivalent:*

i. *P is $l.s.c.$;*

ii. *the mapping $t \longrightarrow d(x, P(t))$ is $u.s.c.$ for every x.*

PROOF. $i. \Longrightarrow ii.$

Take arbitrary (t_0, x_0) and $t_\alpha \longrightarrow t_0$. We have to show that condition $d(x_0, P(t_\alpha)) \longrightarrow a$ implie that $a \leq d(x_0, P(t_0))$.

Fix $\varepsilon > 0$ and let $\overline{x} \in P(t_0)$ be such an element that

$$d(x_0, \overline{x}) \leq d(x_0, P(t_0)) + \varepsilon.$$

By the $l.s.c.$ there exist $x_\alpha \in P(t_\alpha)$ such that $x_\alpha \longrightarrow \overline{x}$. Then

$$d(x_0, P(t_\alpha)) \leq d(x_0, x_\alpha) \leq d(\overline{x}, x_\alpha) + d(x_0, P(t_0)) + \varepsilon.$$

Passing to the limit with α and $\varepsilon \longrightarrow 0$ we obtain

$$a \leq d(x_0, P(t_0)),$$

what was to be proved.

$ii. \Longrightarrow i.$

Take arbitrary $(t_0, x_0) \in grP$ and let $t_\alpha \longrightarrow t_0$. Then

$$\limsup d(x_0, P(t_\alpha)) \leq d(x_0, P(t_0)) = 0$$

and hence

$$\lim d(x_0, P(t_\alpha)) = 0.$$

Therefore, for every integer $k \geq 1$ there exists α_k such that for $\alpha \succeq \alpha_k$

$$d(x_0, P(t_\alpha)) < \frac{1}{k}.$$

We may assume that

$$\alpha_1 \preceq \alpha_2 \preceq \dots \preceq \alpha_k \preceq \alpha_{k+1} \preceq \dots.$$

There exist $x_{\alpha,k} \in P(t_\alpha)$ such that $d(x_0, x_{\alpha,k}) < \frac{1}{k}$.

Define $x_\alpha \in P(t_\alpha)$ by taking $x_\alpha = x_{\alpha,k}$ for $\alpha_k \preceq \alpha \prec \alpha_{k+1}$. Then $d(x_0, x_\alpha) < \frac{1}{k}$ for $\alpha_k \preceq \alpha \prec \alpha_{k+1}$, what means that $x_\alpha \longrightarrow x_0$. This completes the proof. □

In a similar way we can prove the following:

PROPOSITION 29. *For a multifunction $P : T \longrightarrow N(X)$ the following conditions are equivalent:*

i. P is u.s.c.;

ii. the mapping $t \longrightarrow d(x, P(t))$ is l.s.c. for every $x \in X$.

We leave the proof for the reader.

A special case of multivalued mappings establish multifunctions having the values in $L^p(T, X)$. The examinations of such mapping can be sometimes reduced to the case of $p = 1$. For this purpose we use so called Mazur transformation

$$\omega_p : L^p(T, X) \longrightarrow L^1(T, X)$$

given by

(4.4)
$$\omega_p(u) = \begin{cases} |u|^q u & \text{if } p > 1 \\ u & \text{if } p = 1 \end{cases},$$

where

$$\frac{1}{p} + \frac{1}{q} = 1.$$

The Mazur transformation is continuous and surjective. It also establishes the following relation:

PROPOSITION 30. *Let S be a topological space. Consider a mapping $P : S \longrightarrow N(L^p(T, X))$ and let $\omega_p(P) : S \longrightarrow N(L^1(T, X))$ be given by*

$$\omega_p(P)(s) = \omega_p\{P(s)\}.$$

Then we have the following characterizations:

i. P is u.s.c. iff $\omega_p(P) : S \longrightarrow N(L^1)$ is u.s.c.;

ii. P is l.s.c. iff $\omega_p(P) : S \longrightarrow N(L^1)$ is l.s.c..

We leave the proof as an exercise.

3. Upper and lower semicontinuity in Banach spaces

Upper and lower semicontinuity in a reflexive Banach space X can also be characterized in terms of the support functions.

For given $P : T \longrightarrow N(X)$ the support function of each $P(t)$ we shall denote by

$$c_P(t, x^*) = \sup\{\langle x^*, x \rangle : x \in P(t)\}.$$

We begin with the following

PROPOSITION 31. *Let* $P : T \longrightarrow b(X)$ *be an u.s.c. mapping. Then the support function* $c_P : T \times B^* \longrightarrow \overline{R}$ *is u.s.c. within the strong topology in* B^*. *Moreover, for* $P : T \longrightarrow cc(Z)$, *where* $Z \in cc(X)$, *the function* c_P *is u.s.c. within the weak* topology in* B^*. *And vice versa, if for* $P : T \longrightarrow cc(Z)$, *where* $Z \in cc(X)$, *it's support function* $c_P : T \times B^* \longrightarrow R$ *is u.s.c. within the weak* topology in* B^* *then* P *is u.s.c..*

PROOF. For the first part it is sufficient to show that whenever

$$t_\alpha \longrightarrow t_0, \quad x_\alpha^* \longrightarrow x_0^* \quad strongly \quad and \quad c_P(t_\alpha, x_\alpha^*) \longrightarrow \gamma$$

then

(4.5) $$\gamma \le c_P(t_0, x_0^*).$$

For any $\varepsilon > 0$ consider set $V = \{x : d(x, P(t_0)) < \varepsilon\}$ and observe that it is open and bounded. Then

$$d_H(V, P(t_0)) \le \varepsilon$$

and thus, by the inequality (1.7),

$$|c_P(t_0, x_0^*) - c_V(x_0^*)| \le \varepsilon.$$

Notice that t_0 is an interior point in $P^+(V)$ and hence for "sufficiently large α" we have $t_\alpha \in P^+(V)$, i.e.

$$P(t_\alpha) \subset V.$$

Thus

$$c_P(t_\alpha, x_\alpha^*) \le c_V(x_\alpha^*) \le c_V(x_0^*) + |V| \|x_\alpha^* - x_0^*\| \le$$
$$\le c_P(t_0, x_0^*) + |V| \|x_\alpha^* - x_0^*\| + \varepsilon,$$

where $|V| = \sup\{|x| : x \in V\}$. Hence

$$\gamma \le c_P(t, x^*) + \varepsilon.$$

But $\varepsilon > 0$ was arbitrarily chosen. So passing to the limit with $\varepsilon \longrightarrow 0$ we get (4.5).

For the second part assume that

$$t_\alpha \longrightarrow t_0, \ x_\alpha^* \longrightarrow x_0^* \ \textit{weakly}^* \ \textit{and} \ c_P(t_\alpha, x_\alpha^*) \longrightarrow \gamma.$$

Take $x_\alpha \in P(t_\alpha) \subset Z$ such that

$$c_P(t_\alpha, x_\alpha^*) = \langle x_\alpha^*, x_\alpha \rangle.$$

By compactness we may assume that $x_\alpha \longrightarrow x_0$. Hence

$$\langle x^*, x_0 \rangle = \gamma \leq c_P(t_0, x_0^*).$$

For the third statement assume that c_P is u.s.c. We need to verify that for every $(t_\alpha, x_\alpha) \in gr P$ such that $(t_\alpha, x_\alpha) \longrightarrow (t_0, x_0)$ one has $(t_0, x_0) \in gr P$. To a contrary assume that it can be found a net $(t_\alpha, x_\alpha) \in gr P$ converging to $(t_0, x_0) \notin gr P$. But then $x_0 \notin P(t_0)$ and by the separation theorem we can select $x^* \in B^*$ such that

$$\langle x^*, x_0 \rangle > c_P(t_0, x^*).$$

Thus, for "sufficiently large α", we have

$$\langle x^*, x_\alpha \rangle > c_P(t_\alpha, x^*),$$

what means that $x_\alpha \notin P(t_\alpha)$. Equivalently, $(t_\alpha, x_\alpha) \notin gr P$, a contradiction. $\qquad \square$

In a similar way we shall prove

PROPOSITION 32. *Let* $P : T \longrightarrow cb(X)$ *be an u.s.c. mapping and fix* $t_n \longrightarrow t_0$. *Then the inclusion*

$$\bigcap_{k=1}^{\infty} clco \left\{ \bigcup_{n=k}^{\infty} P(t_n) \right\} \subset cl P(t_0)$$

holds.

PROOF. Fix any $x^* \in B^*$, $\varepsilon > 0$ and take

$$V_\varepsilon = \{x : \langle x^*, x \rangle < c_P(t_0, x^*) + \varepsilon\}.$$

One can easily notice that V_ε is open and convex. Apply the u.s.c. to find k such that for any $n \geq k$ we have

$$c_P(t_n, x^*) < c_P(t_0, x^*) + \varepsilon$$

Therefore for $n \geq k$ we han inclusions

$$P(t_n) \subset V_\varepsilon.$$

Thus

$$\bigcap_{k=1}^{\infty} clco \left\{ \bigcup_{n=k}^{\infty} P(t_n) \right\} \subset cl V_\varepsilon \subset \{x : \langle x^*, x \rangle \leq c_P(t, x^*) + \varepsilon\}$$

and passing to the limit with $\varepsilon \longrightarrow 0$ we get

$$\bigcap_{k=1}^{\infty} clco \left\{ \bigcup_{n=k}^{\infty} P(t_n) \right\} \subset clV_{\varepsilon} \subset \{x : \langle x^*, x \rangle \leq c_P(t, x^*)\}$$

But $x^* \in B^*$ was arbitrarily chosen, so we get the desired inclusion.
$\square$

For *l.s.c.* mapping we have the following

PROPOSITION 33. *Let* $P : T \longrightarrow b(X)$ *be an l.s.c. mapping. Then the support function* $c_P : T \times B^* \longrightarrow R$ *is l.s.c. within the weak* topology in* B^*. *And vice versa, if the support function of* $P : T \longrightarrow cc(K)$, *where* $K \in cc(X)$, *is l.s.c. then* P *is l.s.c..*

PROOF. For the first part it is sufficient to show that whenever $(t_\alpha, x_\alpha^*) \longrightarrow (t_0, x_0^*)$ and $c_p(t_\alpha, x_\alpha^*) \longrightarrow \gamma$ then

(4.6) $$\gamma \geq c_P(t_0, x_0^*).$$

Take any point $x_0 \in P(t_0)$ and choose a net $x_\alpha \in P(t_\alpha)$ such that $x_\alpha \longrightarrow x_0$. Then

$$c_P(t_\alpha, x_\alpha^*) \geq \langle x_\alpha^*, x_\alpha \rangle$$

and hence

$$\gamma \geq \langle x_0^*, x_0 \rangle.$$

But $x_0 \in P(t_0)$ was arbitrarily chosen, so

$$\gamma \geq \sup \{ \langle x_0^*, x \rangle : x \in P(t_0) \} = c_P(t_0, x_0^*).$$

For the opposite statement assume that $P : T \longrightarrow cc(K)$ is not *l.s.c.* at a point t_0. By definition it means that that there is an open V such that $t_0 \in P^-(V)$ is not an interior point in $P^-(V)$. Hence one can find a net $t_\alpha \longrightarrow t_0$ such that $t_\alpha \notin P^-(V)$, i.e. $P(t_\alpha) \cap V = \varnothing$, while $P(t_0) \cap V \neq \varnothing$. Take $x_0 \in P(t_0) \cap V$. Then there exists $r > 0$ such that $\overline{B}(x_0, r) \subset V$ and thus $P(t_\alpha) \cap \overline{B}(x_0, r) = \varnothing$. By the separation theorem for each α there exists $x_\alpha^* \in B^*$ with $\|x_\alpha^*\| = 1$ such that

$$c_P(t_\alpha, x_\alpha^*) \leq \langle x_\alpha^*, x_0 \rangle - r.$$

Without loss of generality we may assume that x_α^* is weakly* convergent to x_0^*. Then

$$c_P(t_0, x_0^*) \leq \liminf c_P(t_\alpha, x_\alpha^*) \leq \langle x_0^*, x_0 \rangle - r,$$

what means that $x_0 \notin P(t_0)$, a contradiction.
$\square$

4. Michael selection theorem

We aregoing to discuss the continuous selection property for a lower semicontinuous multifunction with closed, convex values and some it's consequences. In this section T stands for a paracompact Hausdorff topological space. As we have seen in the Proposition 25 for an arbitrary family $\{p_\alpha\}_{\alpha \in \Lambda}$ of continuous functions from T into a Banach space X the multifunction $P : T \longrightarrow cl(X)$ given by

$$P(t) = cl\{p_\alpha(t) : \alpha \in \Lambda\}$$

is *l.s.c.* Obviously each p_α is a continuous selection of P. From the other hand there are continuous multifunctions which do not possess a continuous selection property.

EXAMPLE 8. *Let $T \subset R^2$ be the closed unit ball. Consider a multi-valued mapping $P : T \longrightarrow cl(T)$ given for $t = (r \cos \varphi_t, r \sin \varphi_t)$ by*

$$P(t) = \{(\cos \varphi, \sin \varphi) : |\varphi + \varphi_t| \leq \pi(1 - r)\}.$$

Then P is continuous but it has no continuous selection.

PROOF. Obviously P is continuous, since it is *l.s.c.* and *u.s.c.* (It is even Lipschitz continuous with constant π). We shall show that it possess no continuous selection property. Assume to a contrary that $p : T \longrightarrow T$ is a continuous selection of P. Hence $|p(t)| = 1$ for every $t \in B$. By the Schauder Theorem the mapping p has a fixed point t_0 and therefore $|t_0| = |p(t_0)| = 1$. Hence $t_0 = p(t_0) = -t_0$, a contradiction. $\qquad \square$

THEOREM 19 (Michael [**157**]). *Assume that $P : T \longrightarrow clco(X)$ is a l.s.c. multifunction. Then P admits a continuous selection.*

PROOF. The proof goes in three steps. In step I we will construct, for any given $\varepsilon > 0$, an $\varepsilon - approximate$ continuous selection $p : T \longrightarrow X$ such that

$$d(p(t), P(t)) < \varepsilon$$

or equivalently

$$R(t) = P(t) \cap B\{p(t), \varepsilon\} \neq \emptyset.$$

Then by the Example 6.7 the multifunction $R : T \longrightarrow co(X)$ is *l.s.c.*.

In step II we construct two sequences: of continuous functions $p_n : T \longrightarrow X$, $n = 1, 2, ...$ and *l.s.c.* multifunctions

$$P_n : T \longrightarrow clco(X), \quad n = 0, 1, 2, ...$$

having for for every $t \in T$ the following properties:

a) $d(p_{n+1}(t), P_n(t)) < \dfrac{1}{2^{n+1}}$;

and
$$b) \ P_{n+1}(t) \subset P_n(t) \subset P(t).$$
In step III we shall show that $p_n \rightrightarrows p$ locally uniformly and thus p is the required selection.

    Step I: Given $\varepsilon > 0$. Fix $t_0 \in T$ and $x_0 \in P(t_0)$ and consider sets
$$V_{t_0,x_0} = P^- B(x_0, \varepsilon) = \{t \in T : P(t) \cap B(x_0, \varepsilon) \neq \emptyset\}.$$
By the *l.s.c.* the family
$$\{V_{t_0,x_0}\}_{t_0 \in T, x_0 \in P(t_0)}$$
is an open covering of the paracompact space T. Therefore there exists a locally finite continuous partition of unity $\{z_\alpha\}_{\alpha \in \Lambda}$ subordinated to $\{V_{t_0,x_0}\}_{t_0 \in T, x_0 \in P(t_0)}$. This means that:

    a) all $z_\alpha : T \longrightarrow I$ are continuous;

    b) for every $t \in T$ the set $\Lambda(t) = \{\alpha \in \Lambda : z_\alpha(t) > 0\}$ is finite and
$$\sum_{\alpha \in \Lambda(t)} z_\alpha(t) = 1;$$

    c) for given $\alpha \in \Lambda$ there exist $t_\alpha \in \Lambda$ and $x_\alpha \in P(t_\alpha)$ such that
$$(4.7) \qquad z_\alpha^{-1}((0,1]) \subset V_{t_\alpha,x_\alpha}.$$
Notice that (4.7) in particular means that if $z_\alpha(t) > 0$ then
$$P(t) \cap B(x_\alpha, \varepsilon) \neq \emptyset.$$
Denote
$$p(t) = \sum_{\alpha \in \Lambda(t)} z_\alpha(t) x_\alpha$$
and observe that p is continuous, since $\{z_\alpha\}_{\alpha \in \Lambda}$ is a locally finite and continuous partition of unity. We shall show that it is a required approximate selection. Fix t and let $\Lambda(t) = \{\alpha_1, ..., \alpha_m\}$. Then $t \in \bigcap_{i=1}^{m} V_{\alpha_i}$, what, in turn, means that
$$p(t) = \sum_{i=1}^{m} z_{\alpha_i}(t) x_{\alpha_i}$$
and for $i = 1, 2, ..., m$ we have
$$P(t) \cap B(x_{\alpha_i}, \varepsilon) \neq \emptyset.$$
Pick $y_i \in P(t) \cap B(x_{\alpha_i}, \varepsilon)$ and take $y = \sum_{i=1}^{m} z_{\alpha_i}(t) y_{\alpha_i}$. Then by the convexity $y \in P(t)$. We shall check that
$$|y - p(t)| < \varepsilon.$$

Indeed, it follows from relations

$$|y - p(t)| = \left| \sum_{i=1}^{m} z_{\alpha_i}(t) x_{\alpha_i} - \sum_{i=1}^{m} z_{\alpha_i}(t) y_i \right| \leq \sum_{i=1}^{m} z_{\alpha_i}(t) |x_{\alpha_i} - y_i| < \varepsilon,$$

what ends *Step I*.

Step II: The construction goes by the induction argument.

For $n = 0$ put $P_0(t) = P(t)$.

By the step I with $\varepsilon = \frac{1}{2^1}$ there is a continuous $p_1 : T \longrightarrow X$ such that $d(p_1(t), P_0(t)) < \frac{1}{2^1}$. Then, by the Example 6, the multifunction $P_1 : T \longrightarrow clco(X)$ given by

$$P_1(t) = cl \left\{ P_0(t) \cap B \left(p_1(t), \frac{1}{2^1} \right) \right\}$$

is *l.s.c.*

Assume that we have already constructed continuous $p_1, ..., p_n : T \longrightarrow X$ and *l.s.c.* multifunctions $P_0, ..., P_{n-1} : T \longrightarrow clco(X)$ such that for every $t \in T$ and for $k = 0, 1, ..., n-1$ one has

$$d(p_{k+1}(t), P_k(t)) < \frac{1}{2^{k+1}}$$

and

$$P_{n-1}(t) \subset P_{n-2}(t) \subset ... \subset P_0(t).$$

Setting

$$P_n(t) = cl \left\{ P_{n-1}(t) \cap B \left(p_n(t), \frac{1}{2^n} \right) \right\}$$

we obtain a *l.s.c.* multifunction. By the *step I* for $\varepsilon = \frac{1}{2^{n+1}}$ there exists a continuous $p_{n+1} : T \longrightarrow X$ such that

$$R(t) = P_n(t) \cap B \left(p_{n+1}(t), \frac{1}{2^{n+1}} \right)$$

is a *l.s.c.* multifunction with nonempty convex values. Take

$$P_{n+1}(t) = cl \left\{ P_n(t) \cap B \left(p_{n+1}(t), \frac{1}{2^{n+1}} \right) \right\}$$

and observe that $P_{n+1} : T \longrightarrow clco(X)$ is *l.s.c.* . One can easily check that

$$P_{n+1}(t) \subset P_n(t) \subset ... \subset P_0(t)$$

and

$$d(p_{n+1}(t), P_n(t)) < \frac{1}{2^{n+1}},$$

what ends the induction step.

$\underline{Step\ III}$: We shall show that $p_n \rightrightarrows p$ and p is the required selection. For any integers $n \geq 0$ and $n \geq 1$ we have $P_{n+k-1}(t) \subset P_n(t)$
and hence

$$d\left(p_{n+k}(t), P_n(t)\right) \leq d\left(p_{n+k}(t), P_{n+k-1}(t)\right) \leq \frac{1}{2^{n+k}}.$$

Therefore

$$|p_{n+k}(t) - p_n(t)| \leq d\left(p_{n+k}(t), P_n(t)\right) + d\left(p_n(t), P_n(t)\right) \leq$$

$$\leq \frac{1}{2^{n+k}} + \frac{1}{2^n} \leq \frac{1}{2^{n-1}},$$

what shows that sequence $\{p_n\}_{n=1}^{\infty}$ of continuous functions satisfies the
uniform Cauchy condition. Thus there is a continuous p such that
$p_n \rightrightarrows p$. Such p is the required selection since, by (8), we have

$$d\left(p(t), P(t)\right) = \lim_{n \to \infty} d\left(p_n(t), P_0(t)\right) = 0.$$

This completes the proof.   □

REMARK 7. *a) actually the Michael Selection Theorem says more
then we have presented. Original Michael's formulation is the following:*

THEOREM 20. *A Hausdorff topological space is paracompact if and
only if every separable Banach space X possess the property that any
l.s.c. multifunction $P : T \longrightarrow clcoX$ admits a continuous selection.*

We have cited its formulation only from the historical point of view,
since for our purpose just given version is sufficient. Also in many
applications it is used the version as presented.

*b) In the above proof we did not take any advantage of the separa-
bility of X. So the Michael Selection Theorem holds also for arbitrary
Banach space.*

The previous result gives also an answer on a question how many
continuous selections posses a *l.s.c.* multifunction that $P : T \longrightarrow$
$clco\,(X)$. Especially whether for given $t_0 \in T$ a *l.s.c.* multifunction
that $P : T \longrightarrow clco\,(X)$ admits a continuous selection asumming a
prescribed value $x_0 \in P(t_0)$.

PROPOSITION 34. *Consider a l.s.c. multifunction $P : T \longrightarrow clco\,(X)$
and let $T_0 \subset T$ be a given closed subset. Then each continuous selection $p : T_0 \longrightarrow X$ of P restricted to T_0 can be extended to a continuous
selection of P on the whole T. In particular, for any given $t_0 \in T$ and
$x_0 \in P(t_0)$, there exists a continuous selection p_{t_0,x_0} of P such that*

$$p_{t_0,x_0}(t_0) = x_0.$$

PROOF. Consider a mapping $P_0 : T \longrightarrow clco(X)$ given by

$$P_0(t) = \left\{ \begin{array}{ll} \{p(t)\}, & for \quad t \in T_0, \\ P(t) & for \quad t \notin T_0, \end{array} \right\}$$

and observe that P_0 is $l.s.c.$. Therefore it admits a continuous selection and it can be easily noticed that this is a required function. In particular taking $T_0 = \{t_0\}$ we can extend $p_{t_0,x_0}(t_0) = x_0 \in P(t_0)$ to a continuous selection of P. □

THEOREM 21. *Let $P : T \longrightarrow clco(X)$ be a given multifunction. Then P is l.s.c. iff there exist continuous $p_\alpha : T \longrightarrow X$ such that for every $t \in T$*

(4.8) $$P(t) = cl\{p_\alpha(t) : \alpha \in \Lambda\},$$

Additionally, if both T and X are separable then we may assume the representation 4.8 is countable.

PROOF. By the Corollary 34 for any $l.s.c.$ mapping $P : T \longrightarrow clco(X)$ we have

$$P(t) = cl\{p_{t_0,x_0}(t) : t_0 \in T, \ x_0 \in P(t_0)\},$$

what gives (4.8). On the other hand, if (4.8) holds then by the Proposition 25 yields the $l.s.c.$ of P.

A countable representation for separable T and X can be constructed in the following way. Let $\{x_n\}_{n=1}^\infty$ be a dense subset in X. For every $n = 1, 2, \ldots$ consider open sets

$$V_{n,k} = P^- B\left(x_n, \frac{1}{k}\right) = \left\{t : P(t) \cap B\left(x_n, \frac{1}{k}\right) \neq \emptyset\right\}.$$

By the choice of $\{x_n\}_{n=1}^\infty$ we have

$$T = \bigcup_{n,k=1}^\infty V_{n,k}.$$

But in a metrizable space any open set can be decomposed into a countable union of closed sets. Therefore for every $n, k = 1, 2, \ldots$ there exist closed $F_{n,k,m}$, $m = 1, 2, \ldots$, that

$$V_{n,k} = \bigcup_{m=1}^\infty F_{n,k,m}.$$

Consider multifunctions

(4.9) $$P_{n,k,m}(t) = \left\{ \begin{array}{ll} cl\{P(t) \cap B\left(x_n, \frac{1}{k}\right)\} & for \quad t \in F_{n,k,m} \\ P(t) & for \quad t \notin F_{n,k,m} \end{array} \right.$$

and observe that each $P_{n,k,m} : T \longrightarrow clcoX$ is *l.s.c.*. Therefore, by the Michael Selection Theorem, it admits a continuous selection $p_{n,k,m} : T \longrightarrow X$. Obviously each $p_{n,k,m}$ is also a selection of P. We shall show that for every $t \in T$

$$(4.10) \qquad P(t) = cl \{p_{n,k,m}(t) : n, k, m = 1, 2, ...\} .$$

Denote the right-hand side of (4.9) by $R(t)$. Since $R(t) \subset P(t)$ is straightforward we need to show that $P(t) \subset R(t)$. To see this let us fix $t \in T$, $x \in P(t)$ and $k = 1, 2, ...$. Choose n in such way that $x \in P(t) \cap B\left(x_n, \frac{1}{k}\right)$. Then $t \in V_{n,k}$. Thus there exists m that $t \in F_{n,k,m}$ and so $x \in P_{n,k,m}(t)$. Hence for the continuous selection $p_{n,k,m}$ we have

$$|p_{n,k,m}(t) - x_n| \leq \frac{1}{k}$$

and therefore

$$|p_{n,k,m}(t) - x| < \frac{2}{k}.$$

The latter means nothing else then $d(x, R(t)) < \frac{2}{k}$. But k was chosen arbitrarily, so it implies that $x \in R(t)$. By the choise of x we conclude that $P(t) \subset R(t)$, what ends the proof. $\qquad \square$

The the Michael Selection Theorem have also got some important consequences in the real analysis.

COROLLARY 5. *Consider $a, b : T \longrightarrow R$ with $a \leq b$ such that is u.s.c., while $b - l.s.c.$. Assume that for given closed subset $T_0 \subset T$ there is a continuous $c : T_0 \longrightarrow R$ satisfying, for $t \in T_0$, the relation*

$$(4.11) \qquad a \leq p(t) \leq b.$$

Then p can be continuously extended on the whole T preserving (4.11). In particular, for any given $t_0 \in T$ and $x_0 \in P(t_0)$ there exists a continuous selection p_{t_0,x_0} of P such that

$$p_{t_0,x_0}(t_0) = x_0.$$

PROOF. It is a consequence of the Corollary 34 applied for $P(t) = [a(t), b(t)]$ since then $p(t) \in P(t)$ for $t \in T_0$. $\qquad \square$

The Theorem 21 has a counterpart in the real analysis as well. Namely we have

THEOREM 22. *Let $a, b : T \longrightarrow R$ be given mappings. Then:*
i. a is l.s.c. iff there are continuous $a_\alpha : T \longrightarrow R$, $\alpha \in \Lambda$, that for every $t \in T$

$$(4.12) \qquad a(t) = \sup_{\alpha \in \Lambda} a_\alpha(t).$$

Similarly,

  ii. *b is u.s.c. iff it admits a representation*

(4.13)
$$b(t) = \inf_{\alpha \in \Lambda} b_\alpha(t)$$

with continuous $b_\alpha : T \longrightarrow R$, $\alpha \in \Lambda$. *Moreover, for separable metric space* (T, d) *one can require that representations (4.12) and (4.13) are countable.*

PROOF. Since a is *u.s.c.* iff $-a$ is *l.s.c.* therefore we can deal only with the *l.s.c* case. Now our statement is a translation of the the formula (4.8) to $P(t) = (-\infty, a(t)]$. ∎

In thethe Michael Selection Theorem all assumptions are essential. The *Example 8* shows that, in general, we can not get rid of the convexity, clodedness and the *l.s.c.* assumptions. However, in certain situations we can prove the existence of continuous selections with no closedness or convexity assumptions.

PROPOSITION 35. *Let* $P : T \longrightarrow clco(X)$ *be a l.s.c. multifunction and assume that we have given continuous functions* $c : T \longrightarrow X$ *and* $r : T \longrightarrow R^+$ *such that for every* $t \in T$ *the set*

$$R(t) = P(t) \cap B(c(t), r(t)) \neq \emptyset.$$

Then $R : T \longrightarrow co(X)$ *admits a continuous selection.*

PROOF. Using the Proposition 26 we may assume that

$$c \equiv 0 \quad and \quad r \equiv 1.$$

Fix $t_0 \in T$ and $x_0 \in R(t_0) = P(t_0) \cap B(0, 1)$. By the Corollary 21 the multifunction P admits such continuous selection p_{t_0, x_0} that

$$p_{t_0, x_0}(t_0) = x_0.$$

Consider sets

$$V_{t_0, x_0} = \{t \in T : |p_{t_0, x_0}(t)| < 1\}$$

and observe that V_{t_0, x_0} is an open neighbourhood of t_0. Moreover, by the *l.s.c.*, the family $\{V_{t_0, x_0}\}_{t_0 \in T, x_0 \in P(t_0)}$ form an open covering of the paracompact space T. Therefore there exists a locally finite continuous partition of unity $\{z_\alpha\}_{\alpha \in \Lambda}$ subordinated to $\{V_{t_0, x_0}\}_{t_0 \in T, x_0 \in P(t_0)}$. Denote by

$$p(t) = \sum_{\alpha \in \Lambda} z_\alpha(t) p_{t_\alpha, x_\alpha}(t).$$

We claim that p is a required selection. Indeed, p is continuous selection of P, since such are $p_{t_\alpha, x_\alpha}(t)$ and $\{z_\alpha\}_{\alpha \in \Lambda}$ is locally finite and continuous partition of unity. Since easily $|p(t)| < 1$ then the proof is completed. ∎

CHAPTER 5

Measurable multifunctions

1. Definitions and properties

Let T be a set with a $\sigma - field$ Σ of subsets of T and let X and Y be topological spaces.

DEFINITION 14. *A multifunction $P : T \longrightarrow N(X)$ is said to be $\Sigma-measurable$ (or simply measurable) iff for every open $U \subset X$ the set $P^-(U) \in \Sigma$.*

EXAMPLE 9. *A multifunction $P(t) = \{p(t)\}$ with $p : T \longrightarrow X$ is measurable iff p is measurable.*

Other examples are presented below. We also give some characterizations on measurable multifunctions and their properties. We begin with certain equivalence conditions.

PROPOSITION 36. *The following conditions are equivalent:*
(i) $P : T \longrightarrow N(X)$ is measurable;
(ii) for every closed $F \subset X$ the set $P^+(F) \in \Sigma$.
If (X,d) is a separable metric space then both conditions are equivalent to:
(iii) for every open ball $B(x,r)$ the set $P^-\{B(x,r)\} \in \Sigma$.

PROOF. The equivalence (ii) and (iii) can be concluded from the fact that in a separable metrizable space any open set can be decomposed into a countable union of closed sets. The equivalence (i) and (iii) follows from (3.1) and the fact that in a a separable metric space any open $U \subset X$ can be represented in the form $U = \bigcup_{n=1}^{\infty} B(x_n, r_n)$. $\square$

REMARK 8. *One can notice that, by (3.3), a multifunction $P : T \longrightarrow N(X)$ is measurable iff $\overline{P} : T \longrightarrow cl(X)$, given by $\overline{P}(t) = clP(t)$, is measurable.*

In a metrizable space (X, d) the measurability of a multivalued mapping can also be express in terms of the distance d. Namely we have:

PROPOSITION 37. *A multifunction $P : T \longrightarrow N(X)$ is $\Sigma-$measurable iff it satisfies the following two conditions:*
    i. the mapping $t \longrightarrow d(x, P(t))$ is $\Sigma - measurable$ for all $x \in X$;
    ii. the mapping $x \longrightarrow d(x, P(t))$ is continuous for every $t \in T$.
If $\Sigma = \mathcal{L}$ then condition (ii) can be formulated as follows:
    ii'. the mapping $x \longrightarrow d(x, P(t))$ is continuous for a.e. $t \in T$.

PROOF. It follows from the Proposition 36 and an observation that for arbitrary $r > 0$ and $x \in X$ one has $d(x, P(t)) < r \iff t \in P^-(B(x,r))$. $\square$

Measurable multifunctions form a reacher structure then single-valued mappings. Below we present their some useful properties.

PROPOSITION 38. *Let $r : T \longrightarrow R$, $p_n : T \longrightarrow X$ and $P_n : T \longrightarrow N(X)$ be given measurable mappings for $n = 0, 1, 2, ...$ and assume that $\omega : X \longrightarrow Y$ is continuous. Then formulas*

$$P(t) = \bigcap_{n=0}^{\infty} P_n(t),$$

$$S(t) = \bigcup_{n=0}^{\infty} P_n(t),$$

$$C(t) = cl\{p_n(t) : n = 0, 1, 2, ...\},$$

and

$$\omega(P)(t) = \omega\{P(t)\}$$

define measurable multivalued mappings.

PROOF. The measurability of P and Q is a consequence of (3.1) and (3.2). The measurability of C follows from the observation that

$$C(t) = cl\left\{\bigcup_{n=0}^{\infty} R_n(t)\right\}, \quad where \quad R_n(t) = \{p_n(t)\}.$$

Finally, the measurability of $\omega(P)$ follows from (3.4). $\square$

Now we are going to discuss the measurability property for multifunctions assuming values in a Banach space X.

PROPOSITION 39. *Given arbitrary $x^* \in X^*\backslash\{0\}$, $r : T \longrightarrow R$, $p : T \longrightarrow X$ and measurable mapping $P : T \longrightarrow N(X)$. Let us consider multifunctions:*

$$P(t) = B(p(t), r(t)),$$
$$Q(t) = P(t) + p(t),$$
$$H_1(t) = \{x : \langle x^*, x \rangle > r(t)\},$$

$$H_2(t) = \{x : \langle x^*, x \rangle < r(t)\},$$
$$H_3(t) = \{x : \langle x^*, x \rangle \geq r(t)\},$$
$$H_4(t) = \{x : \langle x^*, x \rangle \leq r(t)\}$$

and

$$H_5(t) = \{x : \langle x^*, x \rangle = r(t)\}$$

Then each of the mappings $P, H_1, ..., H_5$ is measurable iff $r : T \longrightarrow R$ is measurable, while Q iff $p : T \longrightarrow X$ is measurable

PROOF. a) The desired equivalence for P and r follows from the Proposition 36 and an observation that

$$P^- \{B(a, \varrho)\} \Longleftrightarrow \begin{cases} \{t : |x_0 - a| < r(t) + \varrho\} & for \ a \neq x_0 \\ \{t : r(t) > 0\} & for \ a = x_0 \end{cases}.$$

The measurability of Q we conclude from the Proposition 37 applying an easy to derive formula

$$d(x, Q(t)) = d(x - p(t), P(t)).$$

b) It sufficies to show only the measurability of H_1, since for H_2 we proceed in the same way, while the measurability of H_3, H_4 and H_5 are consequences of the relations

$$H_3(t) = cl\{H_1(t)\}, \quad H_4(t) = cl\{H_2(t)\} \quad and \quad H_5(t) = H_3(t) \cap H_4(t).$$

Take $0 \neq a \in X$ such that $\langle x^*, a \rangle = 1$ and denote $V = \{x : \langle x^*, x \rangle < 0\}$. It can be easily checked that

$$H_1(t) = -V + r(t) a.$$

If r is measurable then, by the case a), H_1 is as well. On the other hand if H_1 is measurable then for every q the set

$$T_1 = \{t : qa \in clH_1(t)\} = \{t : d(qa, clH_1(t)) = 0\} \in \mathcal{L}.$$

But

$$T_1 = \{t : (r(t) - q) a \in clV\} = \{t : r(t) \leq q\},$$

thus r is measurable, what ends the proof. $\square$

EXAMPLE 10. *A multifunction $I(t) = [a(t), b(t)]$ is measurable iff the end-point mappings $a = a(\cdot)$ and $b = b(\cdot)$ are measurable.*

PROOF. $\Longrightarrow$ The measurability of a we deduce from a fact that

$$a(t) < x \Longleftrightarrow t \in I^- \{(-\infty, x)\}.$$

Similarly for b we have

$$b(t) > x \Longleftrightarrow t \in I^- \{(x, \infty)\}.$$

$\Longleftarrow$ It follows from the Proposition 39 and an identity

$$I(t) = \frac{a(t) + b(t)}{2} + \overline{B}\left(0, \frac{b(t) - a(t)}{2}\right).$$

$\square$

2. Measurable selections

A mapping $p : T \longrightarrow X$ is called a selection of $P : T \longrightarrow N(X)$ iff

$$p(t) \in P(t) \quad for \quad every \quad t \in T.$$

If p is measurable (continuous) then we call it a measurable (continuous) selection of P.

In the case of a compact Hausdorff space T with a $\sigma - field$ $\mathcal{L}$ of Lebesgue measurable sets given by a Radon measure μ by a measurable selection we also mean a mapping $p : T \longrightarrow X$ that

$$p(t) \in P(t) \quad for \quad almost \quad every \quad (a.e.) \quad t \in T.$$

The fundamental result in the theory belongs to Kuratowski and Ryll-Nardzewski [**139**] and we formulate it as follows:

THEOREM 23 (Kuratowski & Ryll-Nardzewski). *Assume that* $P :$ $T \longrightarrow cl(X)$ *is a measurable multifunction. Then* P *admits a measurable selection.*

PROOF. Conceptually the proof goes in a similar way to that of the Michael Theorem and is done again in three steps. While steps II and III are almost identical, so step I contains similar ideas but it is based on different arguments.

Passing, if necessary, to an equivalent metric we may assume that diameter

$$\delta(X) = \sup\{d(x,y) : x,y \in X\} \leq 1.$$

Step 1. for any given $\varepsilon > 0$ we will find a measurable $p : T \longrightarrow X$ such that

(a) $R(t) = P(t) \cap B\{p(t), \varepsilon\} \neq \emptyset$
    and
(b) multifunction R is measurable;

Step 2. we construct two sequences: of measurable functions $p_n : T \longrightarrow X$ and measurable multifunctions $P_n : T \longrightarrow cl(X)$ having for every $t \in T$ and $n = 1, 2, \dots$ the following properties:

(c) $d(p_n(t), P_n(t)) \leq \frac{1}{2^n}$;
    and
(d) $P_{n+1}(t) \subset P_n(t) \subset P(t)$.

Step 3.   we shall show that $p_n \rightrightarrows p$ and p occures to be a required selection.

Proof of step 1:   Let $\{x_n\}_{n=1}^{\infty}$ be a dense subset in X. Fix $0 < \varepsilon \leq 1$ and let

$$T_n = P_n^- \{B(x_n, \varepsilon)\} = \{t : P(t) \cap B(x_n, \varepsilon) \neq \emptyset\}.$$

Observe that by measurability of P the sets $T_n \in \Sigma$. Moreover, by the density of $\{x_n\}_{n=1}^{\infty}$, there is

$$\bigcup_{n=1}^{\infty} T_n = T.$$

Consider the sets A_n given by:

$$A_1 = T_1, \quad A_n = T_n \setminus \bigcup_{k=1}^{n-1} A_k \quad for \ n \geq 2.$$

Obviously $A_n \subset T_n$ are disjoint measurable sets with

(5.1)
$$\bigcup_{n=1}^{\infty} A_n = T.$$

Define $p = \sum_{n=1}^{\infty} x_n \chi_{A_n}$, where χ_A stands for the characteristic functions of A and observe that p is measurable. To see that (a) holds notice that, by (5.1), for any $t \in T$ there is n such that $t \in A_n \subset T_n$ and then $p(t) = x_n$. Thus

$$P(t) \cap B(p(t), \varepsilon) \neq \emptyset.$$

The property (b) we conclude from the relations

$$R^-(U) = \{t : R(t) \cap U \neq \emptyset\} = \{t : P(t) \cap U \cap B\{p(t), \varepsilon\} \neq \emptyset\} =$$

$$= \bigcup_{n=1}^{\infty} \{t \in A_n : P(t) \cap U \cap B(x_n, \varepsilon) \neq \emptyset\} = \bigcup_{n=1}^{\infty} A_n \cap P^-(B(x_n, \varepsilon)) \cap U.$$

Proof of step 2:   The construction goes by induction argument.

For $n = 0$ fix $x_0 \in X$, take $P_0(t) = P(t)$ and put $p_0(t) = x_0$. Then p_0 is measurable and b. holds since

$$d(p_0(t), P_0(t)) \leq \delta(X) \leq \frac{1}{2^0}.$$

Assume that we have already constructed, for $k = 0, 1, ..., n-1$, measurable p_k and $P_k : T \longrightarrow cl(X)$ such that for every $t \in T$

$$d(p_k(t), P_k(t)) \leq \frac{1}{2^k}$$

and
$$P_{n-1}(t) \subset P_{n-2}(t) \subset \ldots \subset P_0(t).$$
By the *Step 1*, for $\varepsilon = \frac{1}{2^n}$, there exists measurable $p_n : T \longrightarrow X$ such that
$$R(t) = P_{n-1}(t) \cap B\left(p_n(t), \frac{1}{2^n}\right)$$
is a measurable multifunction with nonempty values.

Consider $P_n(t) = P_{n-1}(t) \cap \overline{B}\left(p_n(t), \frac{1}{2^n}\right)$ and observe that $P_n : T \longrightarrow cl(X)$ is measurable. One can easily check that
$$P_n(t) \subset P_{n-1}(t) \subset \ldots \subset P_0(t)$$
and
$$d(p_n(t), P_n(t)) \leq \frac{1}{2^n},$$
what ends the induction step.

Proof of step 3:   We shall show that $p_n \rightrightarrows p$ and p occure to be the required selection. For any integers $n \geq 0$ and $k \geq 1$ we have $P_{n+k}(t) \subset P_n(t)$ and hence
$$(5.2) \qquad d(p_{n+k}(t), P_n(t)) \leq d(p_{n+k}(t), P_{n+k}(t)) \leq \frac{1}{2^{n+k}}.$$
Then
$$(5.3) \quad d(p_{n+k}(t), p_n(t)) \leq d(p_{n+k}(t), P_n(t)) + d(p_n(t), P_n(t)) \leq$$
$$\leq \frac{1}{2^{n+k}} + \frac{1}{2^n} \leq \frac{1}{2^{n-1}},$$
what shows that the sequence $\{p_n\}_{n=1}^{\infty}$ of measurable functions satisfies the uniform Cauchy condition. Therefore there is a measurable p such that $p_n \rightrightarrows p$. Such p is the required selection since, by (5.2), we have the estimate
$$d(p_k(t), P_0(t)) \leq \frac{1}{2^k}.$$
$\square$

The Kuratowski and Ryll-Nardzewski Selection Theorem 23 [**139**] was published in 1965. Shortly after it had appeared a paper by Castaing [**42**] which is complementary to the Theorem 23. On its basis lies the previously made in the Proposition 38 observation that for a sequence of measurable functions $p_n : T \longrightarrow X$, $n = 1, 2, \ldots$ the multifunction
$$(5.4) \qquad P(t) = cl\{p_n(t) : n = 1, 2, \ldots\}$$
is measurable.

DEFINITION 15. *Any family of measurable functions* $p_n : T \longrightarrow X$, $n = 1, 2, \ldots$ *such that (5.4) holds we call a Castaing representation of* $P : T \longrightarrow N(X)$.

A situation in the above definition can be reverse. Namely, we have the following characterizations of measurable multifunctions through measurable selections:

THEOREM 24. *A multifunction* $P : T \longrightarrow cl(X)$ *is measurable iff it admits a Castaing representation by a countable family of measurable selections*

$$p_n : T \longrightarrow X, \quad n = 1, 2, \ldots.$$

PROOF. In view of the Proposition 38 it is enough to demonstrate $\Longrightarrow$.

Let $\{x_n\}_{n=1}^{\infty}$ be a dense subset of X. For every $k, n = 1, 2, \ldots$ consider sets

$$T_{n,k} = P^{-} B\left(x_n, \frac{1}{k}\right)$$

and observe that they are measurable and

$$T = \bigcup_{n,k=1}^{\infty} T_{n,k}.$$

Consider multifunctions $P_{n,k} : T \longrightarrow cl(X)$ given by

$$P_{n,k}(t) = \begin{cases} P(t) & \text{for } t \notin T_{n,k} \\ cl\{P(t) \cap B(x_n, \frac{1}{k})\} & \text{for } t \in T_{n,k} \end{cases}$$

and notice that each $P_{n,k} : T \longrightarrow cl(X)$ is measurable. Therefore, by the Theorem 23, it admits a measurable selection $p_{n,k}$. Renumerating $p_{n,k}$ we obtain a countable set of measurable functions which is a Castaing representation of P, what has to be proved. $\square$

3. Convex measurable multifunctions

3.1. A Banach space case. The measurability of multifunctions in a Banach space X can also be examined with the use of support functions. Recall that by the support function for any $A \in N(X)$ it is ussualy meant the correspondence $c_A : X^* \longrightarrow R\overline{R} = R \cup \{\infty\}$ given by

$$c_A(x^*) = \sup\{\langle x^*, x \rangle : x \in A\}.$$

Our first observation is the following:

THEOREM 25. *Let $P : T \longrightarrow N(X)$ be a measurable multivalued mapping. Then the function $c_P : T \times X^* \longrightarrow R \cup \{\infty\}$ given by $c_P(t, x^*) = \sup\{\langle x^*, x \rangle : x \in P(t)\}$ is $\Sigma \otimes \mathcal{B} -$ measurable. Moreover, if P is measurably bounded then c_P is Lipschitz in x^* and therefore $\Sigma \otimes \mathcal{B}-$measurable.*

PROOF. The $\Sigma \otimes \mathcal{B} - measurability$ follows from the fact that for a Castaing representation of

$$clP(t) = cl\{p_n(t) : n = 1, 2, ...\}$$

we have

$$c_P(t, x^*) = c_{clP}(t, x^*) = \sup_n \{\langle x^*, p_n(t) \rangle\}.$$

To show the second part assume that $P(\cdot)$ is bounded by measurable $p(\cdot)$. Then $x^* \longrightarrow c_P(t, x^*)$ is Lipschitz for every $t \in T$.   □

The situation described in the above theorem can be reversed for $P : T \longrightarrow clco(X)$. Namely we have:

PROPOSITION 40. *Suppose that X^* is separable and consider a multifunction $P : T \longrightarrow clco(X)$. If the support function*

$$c_P(t, x^*) : T \times X^* \longrightarrow R \cup \{\infty\}$$

is $\Sigma \otimes \mathcal{B} - measurable$ then P is measurable.

PROOF. Take a dense subset $\{x_n^*\}_{n=1}^{\infty} \subset B^*$. Denote by $c_n(t) = c_P(t, x_n^*)$ and observe that by (1.8) we have

$$P(t) = \bigcap_{n=1}^{\infty} \{x : \langle x_n^*, x \rangle \le c_n(t)\}.$$

But all $c_n : T \longrightarrow R \cup \{\infty\}$ are measurable. Hence P is measurable as well.   □

REMARK 9. *Using the support function argument we can also explain measurability of multivalued mappings considered in Proposition 39 and the Example 10.*
   (a) *Since*

$$P(t) = \overline{B}(x_0, p(t)) = x_0 + \overline{B}(0, p(t))$$

so for a purpose of the Proposition 39 we may assume that $x_0 = 0$. Then the required equivalence is a consequence of a formula $c_P(t, x^*) = r(t)|x^*|$.

(b)   *The conclusions in the Example 10 follows from an easy to derive formula that for* $I(t) = [a(t), b(t)]$ *we have*

$$
c_I(t, x^*) = \begin{cases} b(t)\, x^* & for \quad x^* > 0 \\ -a(t)\, x^* & for \quad x^* < 0 \\ 0 & for \quad x^* = 0 \end{cases}.
$$

3.2. An R^l case. For measurable multifunctions $P : T \longrightarrow cc\left(R^l\right)$ the existence of measurable selections can also be obtained by an use of extremal selections, i.e. selections of the profile $extP(t)$. For this purpose we shall take an advantage of the Theorem 4. For given orthonormal basis $\mathcal{E} \in \Xi$ and every $t \in T$ denote by $e(t, \mathcal{E})$ the lexicographical maximum of $P(t)$.

THEOREM 26 (Olech [**171**]). *For any orthonormal basis $\mathcal{E} \in \Xi$ the function $e(t, \mathcal{E})$ is a measurable selection of P. Moreover*

(5.5) $$ P(t) = \bigcap_{\mathcal{E} \in \Xi} \left\{ x : x \underset{\mathcal{E}}{\le} e(t, \mathcal{E}) : \mathcal{E} \in \Xi \right\} $$

and

$$ extP(t) = \{ e(t, \mathcal{E}) : \mathcal{E} \in \Xi \}. $$

PROOF. We just need to demonstrate, for every given basis $\mathcal{E} = \{\mathbf{e}_1, \mathbf{e}_2, .., \mathbf{e}_l\}$, the measurability of $e(t, \mathcal{E})$. Consider the multifunctions

$$ P_1(t) = \left\{ x \in P(t) : \langle \mathbf{e}_1, x \rangle = c_{P(t)}(\mathbf{e}_1) \right\} $$

and, inductively, for $i = 1, ...$

$$ P_{i+1}(t) = \left\{ x \in P_i(t) : \langle \mathbf{e}_{i+1}, x \rangle = c_{P(t)}(\mathbf{e}_{i+1}) \right\}. $$

Then each P_i is, by the Proposition 39, measurable. Moreover

$$ \dim P_i(t) \le l - i. $$

Therefore

$$ \dim P_l(t) = 0, $$

what means that $P_n(t)$ reduces to just one point. One can easily verify that

$$ P_l(t) = \{ e(t, \mathcal{E}) \}, $$

what yields the measurability of $e(t, \mathcal{E})$.   □

The profile of a convex set may not be closed, however as we have seen, the multifunction

$$ t \longrightarrow extP(t) $$

admits measurable selections. Moreover, complementarily to the Olech's result we have the following measurable version of the Carathéodory Convexity Theorem 2.

THEOREM 27. *Let $P : T \longrightarrow cc\left(R^l\right)$ be a measurable multifunction. Then each measurable selection p of P can be represented as*

$$p(t) = \sum_{i=0}^{l} \lambda_i(t) e_i(t),$$

with measurable $\lambda_i : T \longrightarrow [0,1]$ and $e_i(t) \in extP(t)$ such that

$$\sum_{i=0}^{l} \lambda_i(t) = 1.$$

PROOF. We shall proceed by the induction. For $l = 1$ the assumptions lead to $P(t) = [a(t), b(t)]$ with measurable $a, b : T \longrightarrow R$. For given measurable selection $p(t) \in [a(t), b(t)]$ denote by $A = \{t : a(t) < b(t)\}$ and take

$$\lambda(t) = \left\{ \begin{array}{cc} \frac{p(t) - a(t)}{b(t) - a(t)} & for \ \ t \in A \\ 0 & for \ \ t \notin A \end{array} \right..$$

Then $\lambda : T \longrightarrow [0, 1]$ is measurable and

$$p(t) = \lambda(t) b(t) + (1 - \lambda(t)) a(t).$$

Assume that the theorem holds for all $k \leq l-1$ and consider measurable $P : T \longrightarrow cc\left(R^l\right)$. Without loss of generality we may require that

$$e_l(t) = 0 \in extP(t).$$

Fix a measurable selection p of P and let

$$P_l(t) = \{x \in P(t) : \langle x, p(t) \rangle = c_P(t, p(t))\}.$$

Then P_l is a measurable multifunction with $\dim P_l(t) \leq l - 1$. Denote by

$$A = \{t : p(t) \neq 0\} = \{t : c_P(t, p(t)) > 0\}$$

and consider the function

$$p_l(t) = \left\{ \begin{array}{cc} \frac{c_P(t, p(t))}{|p(t)|^2} p(t) & for \ \ t \in A \\ 0 & for \ \ t \notin A \end{array} \right..$$

One can easily check that p_l is a measurable selection of P_l. By the induction step there exist measurable $e_i(t) \in extP_l(t) \subset extP(t)$ and $\lambda_i : T \longrightarrow [0, 1]$ with $\sum_{i=0}^{l-1} \lambda_i(t) = 1$ such that

$$p_l(t) = \sum_{i=0}^{l-1} \lambda_i(t) e_i(t).$$

Then

$$p(t) = \sum_{i=0}^{l-1} \frac{\lambda_i(t) |p(t)|^2}{c_P(t, p(t))} e_i(t) = \sum_{i=0}^{l} \widetilde{\lambda}_i(t) e_i(t),$$

where

$$\widetilde{\lambda}_i(t) = \begin{cases} \frac{\lambda_i(t)|p(t)|^2}{c_P(t,p(t))} & for \ t \in A \\ 0 & for \ t \notin A \end{cases}, \quad i = 0, 1, ..., l-1$$

and

$$\widetilde{\lambda}_l(t) = \begin{cases} 1 - \frac{|p(t)|^2}{c_P(t,p(t))} & for \ t \in A \\ 1 & for \ t \notin A \end{cases}.$$

Now an observation that all $\widetilde{\lambda}_i : T \longrightarrow [0,1]$ are measurable and $\sum_{i=0}^{l} \widetilde{\lambda}_i(t) = 1$ ends the proof. $\qquad\square$

4. Connection of measurability with u.s.c. and l.s.c.

In what follows T is a complete separable metric space with a $\sigma -$ *field* $\mathcal{L}$ of Lebesgue measurable sets given by a locally finite Radon measure μ, while (X, d) stands for a separable metric space. If it is not specifically stated saying about measurability we mean $\mathcal{L} -$ *measurability*.

PROPOSITION 41 (Castaing [42], Jacobs [118]). *Let $P : T \longrightarrow cl(X)$ be a measurable multivalued mapping. Then for every $\varepsilon > 0$ there exists a compact set $T_\varepsilon \subset T$, with $\mu(T \backslash T_\varepsilon) \leq \varepsilon$ such that P restricted to T_ε has the closed graph.*

PROOF. Fix $\varepsilon > 0$ and let $D = \{z_n\}_{n \in N}$ be a dense subset in X. Then each mapping $d_n(t) = d(z_n, P(t))$ is measurable and by the Lusin Theorem there exist compact sets $T_n \subset T$, with $\mu(T \backslash T_n) \leq \frac{\varepsilon}{2^n}$ such that each d_n restricted to T_n is continuous. Take $T_\varepsilon = \bigcap_{n=1}^{\infty} T_n$ and observe that it is compact with $\mu(T \backslash T_\varepsilon) \leq \varepsilon$ and each d_n restricted to T_ε is continuous. We claim that P restricted to T_ε has the closed graph. Indeed, let $(t_k, x_k) \in grP \cap (T_\varepsilon \times X)$ be a sequence convergent to (t_0, x_0). Fix $\delta > 0$ and take a member z_n from D such that $d(z_n, x_0) < \delta$. Then for sufficiently large $k \geq k(n)$ one has $d(z_n, P(t_k)) \leq d(z_n, x_k) < \delta$. Thus the continuity of d_n yields $d(z_n, P(t_0)) \leq \delta$ and therefore $d(x_0, P(t_0)) < 2\delta$. But δ was arbitrarily chosen. Thus $d(x_0, P(t_0)) = 0$ what means that $(x_0, t_0) \in grP$ and ends the proof. $\qquad\square$

A similar facts holds also for the lower semicontinuity.

PROPOSITION 42. *Let* $P : T \longrightarrow cl\,(X)$ *be a measurable multivalued mapping. Then for every* $\varepsilon > 0$ *there exists a compact set* $T_\varepsilon \subset T$, *with* $\mu\,(T \backslash T_\varepsilon) \leq \varepsilon$ *such that* P *restricted to* T_ε *is l.s.c..*

PROOF. Take a Castaing representation of the mapping P by the measurable functions $p_n : T \longrightarrow X$, $n = 1, 2, \ldots$ i.e.

$$P\,(t) = cl\,\{p_n\,(t)\,, \quad n = 1, 2, \ldots\}\,.$$

By the Lusin Theorem for every $\varepsilon > 0$ there exists a compact set $T_\varepsilon \subset T$, with $\mu\,(T \backslash T_\varepsilon) \leq \varepsilon$, such that each p_n restricted to T_ε is continuous. But this, in view of the Proposition 25, means that P is *l.s.c..*   □

Combining both previous results we have

COROLLARY 6. *Let* $P : T \longrightarrow cl\,(X)$ *be a measurable multivalued mapping. Then for every* $\varepsilon > 0$ *there exists a compact set* $T_\varepsilon \subset T$, *with* $\mu\,(T \backslash T_\varepsilon) \leq \varepsilon$ *such that* P *restricted to* T_ε *is continuous.*

CHAPTER 6

Carathéodory type multifunctions

1. Jointly measurable functions

Assume now that X is a Banach space with separable X^*, $(T, \mathcal{L}, \mu)$ a complete separable metric space with a $\sigma - field$ $\mathcal{L}$ of Lebesgue measurable sets given by a locally finite Radon measure μ and let (S, Σ) a measure space possessing the projection property (1.11) with respect to $(T, \mathcal{L}, \mu)$. We shall identify $f, g \in M(T \times S, X)$ if for every $s \in S$ we have

$$f(., s) = g(., s) \quad a.e. \ in \ T.$$

In further consideration we need the following

PROPOSITION 43. *Let* $f : T \times S \longrightarrow X$ *be such jointly measurable function that there is* $p \in L^1(T)$ *having property that for every* $s \in S$

$$|f(t, s)| \leq p(t) \quad a.e. \ in \ T.$$

Then the function $F : S \longrightarrow X$ *given by*

$$F(s) = \int_T f(t, s) \, \mu(dt)$$

is $\Sigma - measurable.$

PROOF. If $f = \chi_A$ for some $\mathcal{L} \otimes \Sigma - measurable$ set A then for every $s \in S$ the sets

$$A_s = \{t : (t, s) \in A\} = proj_T A \in \mathcal{L}$$

and thus the mapping

$$s \longrightarrow A_s$$

is $\Sigma - measurable$ from S into $\mathcal{L}$. This, in turn, gives the $\Sigma - measurability$ of the function

$$s \longrightarrow \mu(A_s).$$

If $f = \sum\limits_{i \in I} x_i \chi_{A_i}$ is a simple function with $\mathcal{L} \otimes \Sigma - measurable$ sets A_i then

$$\int_T f(t,s) \mu(dt) = \sum_{i \in I} x_i \mu(A_{is})$$

with $A_{is} = \{t : (t,s) \in A_i\} \in \mathcal{L}$ and so it is $\Sigma - measurable$.

Finally, given jointly measurable $f : T \times S \longrightarrow X$ we can pointwise approximate by such simple functions $f_n : T \times S \longrightarrow X$, $n = 1, 2, \ldots$ that for every $n \in N$

$$|f_n(t,s)| \le p(t) \quad a.e..in \ T.$$

Now, by the Lebesgue Dominated Convergence Theorem, we have

$$F(s) = \lim_{n \longrightarrow \infty} \int_T f_n(t,s) \mu(dt).$$

But each $s \longrightarrow \int_T f_n(t,s) \mu(dt)$ is $\Sigma - measurable$, so the same holds for F. This completes the proof. $\qquad\square$

2. Upper and lower Carathéodory type real functions

In what follows we shall additionally assume that (S, d) stands for a separable complete metric space. By joint measurability we further mean $\mathcal{L} \otimes \mathcal{B} - measurability$. Recall that, by the Theorem 6, the space $(S, \mathcal{B})$ possess the projection property with respet to $(T, \mathcal{L}, \mu)$.

As an example of jointly measurable functions we shall consider Carathéodory type functions and some their generalizations.

DEFINITION 16. *A jointly measurable function $f : T \times S \longrightarrow R$ will be called:*

(i) lower Carathéodory type (l.C.) iff for every $t \in T$ the mapping $s \longrightarrow f(t,s)$ is lower semicontinuous (l.s.c.);

(ii) upper Carathéodory type (u.C.) iff for every $t \in T$ the mapping $s \longrightarrow f(t,s)$ is upper semicontinuous (u.s.c.);

(iii) Carathéodory type (or shortly, Carathéodory function) iff is it both u.C. and l.C.

The reader may already have noticed that $f : T \times S \longrightarrow R$ is $u.C.$ type iff $-f$ is $l.C.$

We shall also emphasize that the measurability in t and continuity in s imply joint measurability. But it is not a case for $u.C.$ or $l.C.$ functions, i.e. the measurability in t and $u.s.c.$ or $l.s.c.$ in s do not imply $\mathcal{L} \otimes \Sigma - measurability$.

In the theory of jointly measurable functions an important role play Scorza-Dragoni Theorem which gives their characterization through the joint continuity. A formulation of such type result we proceed by some remarks.

LEMMA 2. *Let* $f : T \times S \longrightarrow R$ *be an lower Carathéodory type function. Then for every* $\varepsilon > 0$ *there exists a compact* T_ε *with* $\mu(T \backslash T_\varepsilon) < \varepsilon$ *such that* $f : T_\varepsilon \times S \longrightarrow R$ *is l.s.c.*

PROOF. Denote the epigraph of f by E, i.e.

$$E = \{(t, s, x) : x \geq f(t, s)\}$$

and notice that, by the Proposition 6, $E \in \mathcal{L} \otimes \mathcal{B}(S \times R)$. Therefore E is the graph of a measurable mapping $t \longrightarrow \mathcal{P}(t)$ given by it's t−sections

$$\mathcal{P}(t) = \{(s, x) : (t, s, x) \in E\} = \{(s, x) : x \geq f(t, s)\}.$$

Moreover, by the *l.s.c.*, each $\mathcal{P}(t)$ is closed. So we have obtained a measurable mapping $\mathcal{P} : T \longrightarrow cl(S \times R)$. Employing the Proposition 41 we conclude that for every $\varepsilon > 0$ there exists of a compact set $T_\varepsilon \subset T$, with $\mu(T \backslash T_\varepsilon) \leq \varepsilon$ such that $\mathcal{P}$ restricted to T_ε has the closed graph. But this means that the set

$$\{(t, s, x) \in T_\varepsilon \times S \times R : x \geq f(t, s)\}$$

is closed. So $f : T_\varepsilon \times S \longrightarrow R$ is *l.s.c.*   □

Using the Lemma we shall provide a characterization of *l.C.* and *u.C.* functions by the use of the Carathéodory type mappings.

THEOREM 28 (Scorza-Dragoni). *Let* $f : T \times S \longrightarrow R$ *be a jointly measurable function. Then the following conditions are equivalent:*

  i.   f is lower Carathéodory type;

  ii.   for every $\varepsilon > 0$ *there exists a compact* T_ε *with* $\mu(T \backslash T_\varepsilon) < \varepsilon$ *such that* $f : T_\varepsilon \times S \longrightarrow R$ *is l.s.c.;*

  iii.   there exist Carathéodory type functions $f_n : T \times S \longrightarrow R$, $n \in N$, *such that*

$$f(t, s) = \sup_{n \in N} f_n(t, s).$$

Similarly, we have the equivalence of

  i'.   f is upper Carathéodory type;

  ii'.   for every $\varepsilon > 0$ *there exists a compact* T_ε *with* $\mu(T \backslash T_\varepsilon) < \varepsilon$ *such that* $f : T_\varepsilon \times S \longrightarrow R$ *is u.s.c.;*

  iii'.   there exist Carathéodory type functions $f_n : T \times S \longrightarrow R$, $n \in N$, *such that*

$$f(t, s) = \inf_{n \in N} f_n(t, s).$$

PROOF. It is enough to demonstrate just the *l.s.c.* case. Notice also that the implication $i. \implies ii.$ is the content of the Lemma, while $iii. \implies i.$ is a consequence of the Theorem 22. So we just need to give an explanation for $ii. \implies iii.$

Take a family of compact sets $T_m \subset T$, $m = 1, 2, \ldots$ with $\mu(T \backslash T_m) < \frac{1}{m}$ such that $f : T_m \times S \longrightarrow R$ is *l.s.c.*. Moreover we may assume that $\{T_m\}$ is an increasing sequence. Denoting

$$T_0 = \bigcup_{m=1}^{\infty} T_m$$

we obtain a set of full measure. For every m the function $f : T_m \times S \longrightarrow R$, by the Theorem 22, admits a representation

$$f(t, s) = \sup_{n \in N} f_{n,m}(t, s),$$

by continuous functions $f_{n,m} : T_m \times S \longrightarrow R$, $n \in N$. Fix n and observe that each $f_{n,m}$ is a continuous selection of

$$P(t, s) = [f(t, s), \infty)$$

restricted to $T_m \times S$. So, by the Proposition 34, it admits a continuous extension, denoted again by $f_{n,m}$, on T preserving for $(t, s) \in T_0 \times S$ the inequality

$$f(t, s) \geq f_{n,m}(t, s).$$

Setting $f_{n,m}(t, s) = 0$ for $(t, s) \in (T \backslash T_0) \times S$ we obtain a required representation. □

REMARK 10. *Originally the classical Scorza-Dragoni Theorem has covered only the case of a Carathéodory type function $f : [a, b] \times R \longrightarrow R$. Just provided and further generalizations belong to many authors, cf. Castaing&Valadier [42], Cellina [49], Jarnik&Kurzweil [120], Rzeżuchowski [211], Rybiński [208], Artstein-Prikry [6] and others.*

3. Carathéodory type functions in Banach spaces

Further results of the Scorza-Dragoni type we shall obtain by passing to mappings of one variable which assume the values in $M(T, X)$. For any Carathéodory type function $f : T \times S \longrightarrow X$ denote by $\mathbf{f}_T \in M(S, M(T, X))$ and $\mathbf{f}_S \in M(T, M(S, X))$ the mappings given by

$$\mathbf{f}_T(s) = f(\cdot, s) \quad and \quad \mathbf{f}_S(t) = f(t, \cdot).$$

In certain situations the mappings $\mathbf{f}_T$ and $\mathbf{f}_S$ may occur to be measurable. We begin with

PROPOSITION 44. *For $f \in M(T \times S, R)$ the following conditions are equivalent:*

(i) *f is a Carathéodory type function;*

(ii) *$\mathbf{f}_S : T \longrightarrow C = C(S, R)$ is $\mathcal{L} - measurable$.*

PROOF. Since the measurability of $\mathbf{f}_S$ easily implies that f is a Carathéodory type function, so we only need to demonstrate that (i) $\Longrightarrow$ (ii). This is slightly complicated task and our arguments we shall split into few cases.

CASE 1. *Assume that $f : T \times S \longrightarrow R$ is uniformly bounded, i. e.*

$$M = \sup \{\|\mathbf{f}_S(t)\|_\infty : t \in T\} < \infty.$$

Observe that then

$$\mathbf{f}_S : T \longrightarrow C = C(S, R).$$

Since C is a Banach space then it is enough to check that for every closed ball $B = B(\varphi, r) \subset C$ the set

$$\mathbf{f}_S^{-1}(B) \in \mathcal{L}.$$

But

$$\mathbf{f}_S^{-1}(B) = \bigcap_{s \in S} \{t : |f(t, s) - \varphi(s)| \leq r\}$$

and therefore

$$T \backslash \mathbf{f}_S^{-1}(B) = \bigcup_{s \in S} \{t : |f(t, s) - \varphi(s)| > r\} = proj_T A,$$

where $A = \{(t, s) : |f(t, s) - \varphi(s)| > r\} \in \mathcal{L} \otimes \mathcal{B}$. Hence $T \backslash \mathbf{f}_S^{-1}(B) \in \mathcal{L}$, what in turn gives $\mathbf{f}_S^{-1}(B) \in \mathcal{L}$.

CASE 2. *Assume now that $f : T \times S \longrightarrow R^+$. Consider functions $f_n(t, s) = \min(f(t, s), n)$, $n \in N$, and observe that each f_n is an uniformly bounded Carathéodory type function. An application of the case a) yields the $\mathcal{L} - measurability$ of each corresponding $\mathbf{f}_n : T \longrightarrow C = C(S, R)$ given by*

$$\mathbf{f}_n(t)(s) = f_n(t, s).$$

Moreover, we have

$$f_0 \leq f_1 \leq \ldots \quad and \quad \lim_{n \to \infty} f_n(t, x) = f(t, x),$$

so, by the Dini Theorem, for every $t \in T$, $\mathbf{f}_n(t) \longrightarrow \mathbf{f}(t)$ in $C(S, R)$. But this gives desired measurability of $\mathbf{f}$.

CASE 3. *An arbitrary Carathéodory type function* $f : T \times S \longrightarrow R$ *can be decomposed as* $f = f_+ - f_-$, *where*

$$f_+ (t,s) = \max \{f(t,s), 0\} \quad and \quad f_- (t,s) = \max \{-f(t,s), 0\}.$$

Thus the second case produces $\mathcal{L} - measurable$ *functions* $\mathbf{f}_+, \mathbf{f}_- : T \longrightarrow C(S, R)$ *given for every* $t \in T$ *by*

$$\mathbf{f}_+ (t)(s) = f_+ (t,s) \quad and \quad \mathbf{f}_- (t)(s) = f_- (t,s), \quad s \in S$$

But this, in turn, gives the masurability of $\mathbf{f} = \mathbf{f}_+ - \mathbf{f}_-$ *and completes the proof.*

$\square$

The previous Lemma and the Lusin Theorem are the main tools needed to establish a version of the Scorza-Dragoni result.

THEOREM 29. *Assume that* (S, d) *is a separable metric space and let* $f : T \times S \longrightarrow X$ *be jointly measurable function. Then the following conditions are equivalent:*

    i. $f : T \times S \longrightarrow X$ *is a Carathéodory type one;*

    ii. $\mathbf{f}_S : T \longrightarrow C = C(S, X)$ *is* $\mathcal{L} - measurable$;

    iii. for every $\varepsilon > 0$ *there exists a compact* T_ε *with* $\mu(T \backslash T_\varepsilon) < \varepsilon$ *such that* $f : T_\varepsilon \times S \longrightarrow X$ *is continuous;*

    iv. for every $\varepsilon > 0$ *there exists a compact* T_ε *with* $\mu(T \backslash T_\varepsilon) < \varepsilon$ *such that* $\mathbf{f}_{T_\varepsilon} : S \longrightarrow C(T_\varepsilon, X)$ *is continuous.*

PROOF. *i.* $\Longrightarrow$ *ii.*

In view of the Proposition 37 it is enough to verify that for every $\varphi \in C(S, X)$ the distance function

$$t \longrightarrow d(\mathbf{f}_S (t), \varphi)$$

is $\mathcal{L} - measurable$. For this purpose observe that

$$d(\mathbf{f}_S (t), \varphi) = \sup_{s \in S} \arctan |f(t,s) - \varphi(s)| = \sup_{n \in N} \arctan |f(t, s_n) - \varphi(s_n)|,$$

where $\{s_n\}_{n \in N}$ is a dense subset in S. But, by assumptions, each function

$$t \longrightarrow \arctan |f(t, s_n) - \varphi(s_n)|$$

is $\mathcal{L} - measurable$ and so is for

$$t \longrightarrow \sup_{n \in N} \arctan |f(t, s_n) - \varphi(s_n)| = d(\mathbf{f}_S (t), \varphi).$$

ii. $\Longrightarrow$ *iii.*

Denote by

$$T_n = \{t : d(\mathbf{f}_S (t), \mathbf{0}) \leq \arctan n\} = \{t : \|\mathbf{f}_S (t)\|_\infty \leq n\}.$$

Then $\{T_n\}_{n\in N}$ is an increasing family of measurable subsets of T such that

$$\bigcup_{n=1}^{\infty} T_n = T$$

and each

$$\mathbf{f}_S : T_n \longrightarrow C(S, X)$$

is $\mathcal{L} - measurable$. Then by the Lusin theorem for every $\varepsilon > 0$ there exist compact $K_n \subset T_n$ with $\mu(T_n \backslash K_n) < \frac{\varepsilon}{2^n}$ such that for every $n \in N$ the mapping $\mathbf{f}_S : K_n \longrightarrow C(S, X)$ is continuous. Taking

$$T_\varepsilon = \bigcap_{n=1}^{\infty} K_n$$

we obtain a compact set satisfying $\mu(T \backslash T_\varepsilon) < \varepsilon$ and such that

$$\mathbf{f}_S : T_\varepsilon \longrightarrow C(S, X)$$

is continuous. One can easily check that then $f : K \times S \longrightarrow X$ continuous.

$iii. \Longrightarrow iv.$

Take such a compact set T_ε that the function $f : T_\varepsilon \times S \longrightarrow X$ is continuous. We need to demonstrate that $\mathbf{f}_{T_\varepsilon} : S \longrightarrow C(T_\varepsilon, X)$ is continuous. For this purpose choose any sequence $s_n \longrightarrow s_0$ and pick $t_n \in T_\varepsilon$ such that

$$\|\mathbf{f}_{T_\varepsilon}(s_n) - \mathbf{f}_{T_\varepsilon}(s_0)\|_\infty =$$
$$= \sup\{|f(t, s_n) - f(t, s_0)| : t \in T_\varepsilon\} = |f(t_n, s_n) - f(t_n, s_0)|.$$

Hence for every subsequence $\{s_{k_n}\}$ there exists a converging subsequence of $\{t_n\}$, say $t_{k_n} \longrightarrow t_0$. Thus

$$\|\mathbf{f}_{T_\varepsilon}(s_{k_n}) - \mathbf{f}_{T_\varepsilon}(s_0)\|_\infty = |f(t_{k_n}, s_{k_n}) - f(t_{k_n}, s_0)| \longrightarrow 0,$$

what means that $\mathbf{f}_{T_\varepsilon}(s_n) \longrightarrow \mathbf{f}_{T_\varepsilon}(s_0)$ in $C(T_\varepsilon, X)$ and shows desired continuity of $\mathbf{f}_{T_\varepsilon}$.

$iv. \Longrightarrow i.$

It is enough to check that for any $x \in X$ the function $g : T \times S \longrightarrow R$ given by

$$g(t, s) = |f(t, s) - x|$$

is a Carathéodory type one. Fix $x \in X$ and $\varepsilon > 0$. For every $n \in N$ pick such a compact K_n that $\mu(T \backslash K_n) < \frac{\varepsilon}{2^n}$ and $\mathbf{f}_{K_n} : S \longrightarrow C(K_n, X)$ is continuous. Thus taking

$$T_\varepsilon = \bigcap_{n=1}^{\infty} K_n$$

we obtain a compact set T_ε having property that $\mu\left(T\backslash T_\varepsilon\right) < \varepsilon$ and $f : T_\varepsilon \times S \longrightarrow X$ as well as $g : T_\varepsilon \times S \longrightarrow R$ are continuous. Hence the Lemma guarrantees that $g : T \times S \longrightarrow R$ is a Carathéodory type function. This completes the proof. $\qquad\square$

4. Jointly measurable multifunctions

The considerations concerning with jointly measurable functions can be preserved for multivalued mappings, but as we will see later there is a need to distinguish between *upper* and *lower* Carathéodory type multivalued mappings.

Let $P : T \times S \longrightarrow cl\left(X\right)$ be an $\mathcal{L} \otimes \Sigma - measurable$. Then P admits a Castaing representation

(6.1) $$P\left(t, s\right) = cl\left\{p_n\left(t, s\right) : n \in N\right\}$$

with $\mathcal{L} \otimes \Sigma - measurable$ functions $p_n : T \times S \longrightarrow X$.

Fix $s_0 \in S$ and $\mathcal{L} - measurable$ selection $p_0\left(\cdot\right)$ of $P\left(\cdot, s_0\right)$. We are wondering whether exist $\mathcal{L} \otimes \Sigma - measurable$ selections of P which are uniformly in $t \in T$ close to p_0. The answer is positive. This situation explains the following

PROPOSITION 45. *For every $\varepsilon > 0$ there exists a jointly measurable selection p of P such for all $t \in T$ we have*

$$d\left(p\left(t, s_0\right), p_0\left(t\right)\right) \leq \varepsilon.$$

PROOF. Fix $\varepsilon > 0$ and by T_n denote $\mathcal{L} - measurable$ sets given by

$$T_n = \left\{t \in T : d\left(p_0\left(t\right), p_n\left(t, s_0\right)\right) \leq \varepsilon\right\}.$$

By 6.1 we have $T = \bigcup\limits_{n=1}^{\infty} T_n$. Consider the $\mathcal{L} - measurable$ sets A_n given by

$$A_1 = T_1, \quad A_n = T_n\backslash\bigcup\limits_{k=1}^{n-1} A_k \quad for\ n \geq 2$$

and define

$$p\left(t, s\right) = \sum\limits_{n=1}^{\infty} \chi_{A_n}\left(t\right) p_n\left(t, s\right).$$

It can be easily checked that p is a required $\mathcal{L} \otimes \Sigma - measurable$ selection of P. $\qquad\square$

LEMMA 3. *Let $P : T \times S \longrightarrow cl\left(X\right)$ be jointly measurable multivalued mapping, while $\varphi : S \longrightarrow X$ a continuous function. Then the set*

$$T_0 = \bigcap\limits_{s \in S} \left\{t : P\left(t, s\right) \cap \left(B\left(\varphi\left(s\right), r\right)\right) \neq \emptyset\right\} \in \mathcal{L}.$$

PROOF. It is a consequence of the Theorem 6. Denote by

$$A = \{(t, s) : P(t, s) \cap (B(\varphi(s), r)) = \emptyset\}$$

and observe that by assumptions

$$A \in \mathcal{L} \otimes \mathcal{B}.$$

Now

$$(6.2) \quad T \backslash T_0 = \bigcup_{s \in S} \{t : P(t, s) \cap (B(\varphi(s), r)) = \emptyset\} = proj_T A \in \mathcal{L},$$

what completes the proof. □

5. Upper and lower Carathéodory type multifunctions

From now assume that S is a Hausdorff topological space.

DEFINITION 17. *We shall call a multifunction* $P : T \times S \longrightarrow cl(X)$ *an upper Carathéodory (u.C.) type multivalued mapping iff it satisfies the following properties:*

i. $P(\cdot, s)$ *is* $\mathcal{L}$ *− measurable for every* $s \in S$;

ii. $P(t, \cdot)$ *is u.s.c. for every* $t \in T$.

Similarly, we shall call a multifunction $P : T \times S \longrightarrow cl(X)$ a *lower* Carathéodory (*l.C.*) type multivalued mapping iff the following conditions hold:

i'. $P(\cdot, \cdot)$ *is* $\mathcal{L} \otimes \mathcal{B}$ *− measurable*;

ii'. $P(t, \cdot)$ *is l.s.c. for every* $t \in T$.

REMARK 11. *The reader may check that conditions i. and ii. imply that* $P(\cdot, \cdot)$ *is* $\mathcal{L} \otimes \mathcal{B}$ *− measurable. However, if we assume that* $P(\cdot, s)$ *is* $\mathcal{L}$*−measurable for every* $s \in S$ *and* $P(t, \cdot)$ *is l.s.c. for every* $t \in T$ *then it fails to imply the joint measurability (see Oxtoby* [**190**]*)*

As for measurable, *l.s.c.* and *u.s.c.* mappings, the property of being an *u.C* or *l.C.* multifunction can be expressed in terms of distance function. Combining the Propositions 37, 30 and 28 we obtain

PROPOSITION 46. *For a jointly measurable multifunction* $P : S \times T \longrightarrow N(X)$ *the following conditions are equivalent:*

i. P *is l.C.;*

ii. *the mapping* $(t, s, x) \longrightarrow d(x, P(t, s))$ *is* $\mathcal{L} \otimes \mathcal{B}(S \times X)$*− measurable in* (t, s, x) *and u.s.c. in* (s, x).

Similarly, we have the equivalence of conditions

i''. P *is u.C.;*

ii''. *the mapping* $(t, s, x) \longrightarrow d(x, P(t, s))$ *is* $\mathcal{L} \otimes \mathcal{B}(S \times X)$ *− measurable in* (t, s, x) *and l.s.c. in* (s, x).

REMARK 12. *Since the mapping* $x \longrightarrow d\left(x, P\left(t, s\right)\right)$ *is continuous then it is enough to check in condition ii. the u.s.c. just for any dense subset* $\{x_n\}_{n \in N}$ *in* X. *A similar fact holds for ii'.*

Now we are ready to give a Scorza-Dragoni type results for $l.C.$ and $u.C.$ mapping. It is formulated as follows:

THEOREM 30 (Fryszkowski, Artstein, Prikry). *A multivalued mapping* $P : T \times S \longrightarrow cl\left(X\right)$ *is an l.C. one iff for every* $\varepsilon > 0$ *there exists a compact set* T_ε *with* $\mu\left(T \backslash T_\varepsilon\right) < \varepsilon$ *such that* $P : T_\varepsilon \times S \longrightarrow cl\left(X\right)$ *is l.s.c.*

PROOF. In view of the Proposition 46 and the Remark 12 it is enough to verify that both conditions are equivalent to a fact that for any dense subset $\{x_n\}_{n \in N}$ each function

$$p_n\left(t, s\right) = d\left(x_n, P\left(t, s\right)\right)$$

is an $u.C.$ one. For this purpose we need to find an universal compact set $T_\varepsilon \subset T$ with $\mu\left(T \backslash T_\varepsilon\right) < \varepsilon$ such that each p_n restricted to $T_\varepsilon \times S$ is $u.s.c..$ But having constructed by the Theorem 29 compact sets $K_n \subset T$ with $\mu\left(T \backslash K_n\right) < \frac{\varepsilon}{2^n}$ such that each p_n restricted to $K_n \times S$ is $u.s.c.$ we can get

$$T_\varepsilon = \bigcap_{n=1}^{\infty} K_n.$$

$\square$

Similarly one can prove the $u.s.c.$ version

THEOREM 31 (Jarnik&Kurzweil [120], Rzeżuchowski [211]). *A multivalued mapping* $P : T \times S \longrightarrow cl\left(X\right)$ *is an u.C. mapping iff for every* $\varepsilon > 0$ *there exists a compact* T_ε *with* $\mu\left(T \backslash T_\varepsilon\right) < \varepsilon$ *such that* P *restricted to* $T_\varepsilon \times S$ *is u.s.c.*

6. Carathéodory type selections in Banach spaces

In what follows we shall assume that X is a Banach space with separable X^*. Consider an $l.C$ multifunction $P : T \times S \longrightarrow clco\left(X\right)$. We already know that for every $s \in S$ the mapping $t \longrightarrow P\left(t, s\right)$ admits a measurable selection while for every $t \in T$ the multifunction $s \longrightarrow P\left(t, s\right)$ possess a continuous selection. It is very natural to pose a question whether P admits a Carathéodory type selection. The answer is positive and it can be obtained combining the Michael and Kuratowski&Ryll-Nardzewski Theorems. This however can be done in two ways. We want to present both of them since they enlighten some new effects. Both approaches use measurable and continuous selections but apply them in an reversed order. The first one is due to Artstein & Prikry and is based on the provided in previous section a Scorza-Dragoni type result. In the second one we consider for any $t \in T$ the set $\mathcal{P}(t)$ of all possible continuous selections, prove the measurability of correspondence $t \longrightarrow \mathcal{P}(t)$ and then apply the Kuratowski & Ryll-Nardzewski Theorem. We would like to say that it is also possible to repeat the Michael and Kuratowski & Ryll-Nardzewski constructions. Such a construction could be based on the Proposition 46. Before we go to details we have to emphasize that the first result is conceptually simpler and gives more general result.

THEOREM 32 (Fryszkowski [**73**], Rybiński [**208**], Artstein&Prikry [**7**]). *Each l.C. multivalued mapping* $P : T \times S \longrightarrow clco(X)$ *admits a Carathéodory type selection.*

PROOF. Take an inreasing family of compact sets $K_m \subset T$, $m = 1, 2, \ldots$ with $\mu(T \backslash K_m) < \frac{1}{m}$ such that $P : K_m \times S \longrightarrow clco(X)$ is l.s.c.. Therefore the Michael Theorem yields the existence of a continuous selection of P restricted to each $K_m \times S$. Starting with a continuous selection p on $K_1 \times S \longrightarrow X$ we can consequtively extend it to a continuous selection on each $K_m \times S$ and finally on

$$T_0 \times S = \left(\bigcup_{m=1}^{\infty} K_m \right) \times S.$$

But $T_0 \subset T$ is a set of full measure. Thus setting $p(t, s) = 0$ for $(t, s) \in (T \backslash K) \times S$ we obtain a required selection. $\qquad \square$

The above result can be enriched with a "Castaing like representation" by Carathéodory type functions. Namely the following theorem holds:

THEOREM 33. *Let* $P : T \times S \longrightarrow clco(X)$ *be a jointly measurable multivalued mapping. Then the following conditions are equivalent:*
   i. f is l.C. type;
   ii. for every $\varepsilon > 0$ *there exists a compact* K *with* $\mu(T \backslash T_\varepsilon) < \varepsilon$ *such that* $P : T_\varepsilon \times S \longrightarrow clco(X)$ *is l.s.c.;*
   iii. there exist Carathéodory type functions $p_n : T \times S \longrightarrow R$, $n \in N$, *such that*

(6.3) $$P(t, s) = cl \{p_n(t, s) : n \in N\}.$$

PROOF. Since *i.* and *ii.* are, by the Theorem 30, equivalent we shall demonstrate implications *ii.* $\Longrightarrow$ *iii.* $\Longrightarrow$ *i.*
   i. $\Longrightarrow$ *iii.*

Take an inreasing family of compact sets $K_m \subset T$, $m = 1, 2, \ldots$ with $\mu\left(T \backslash K_m\right) < \frac{1}{m}$ such that $P : K_m \times S \longrightarrow clco\left(X\right)$ is $l.s.c.$. Then the Theorem 21 gives on every $K_m \times S$ a representation

$$P\left(t, s\right) = cl\left\{p_{n,m}\left(t, s\right) : n \in N\right\}$$

with continuous functions $p_{n,m} : K_m \times S \longrightarrow X$. Fixing m we can continuously extend each $p_{n,m}$ to $T_0 \times S = \left(\bigcup_{m=1}^{\infty} K_m\right) \times S$ obtaining (6.3) on $T_0 \times S$. Now setting $p_{n,m}\left(t, s\right) = 0$ on $\left(T \backslash T_0\right) \times S$ we obtain the required representation.

$ii. \Longrightarrow iii.$

It is an easy conclusion from the Propositions 36 and 26.   □

The second approach to the existence of Carathéodory type selections is valid only on locally compact metrizable spaces. Assume first that (S, d) is a compact metrizable space. In this case $C\left(S, X\right)$ is a separable Banach space what will allow us to apply the Kuratowski & Ryll-Nardzewski Theorem. Let $P : T \times S \longrightarrow clco\left(X\right)$ be a given $l.C.$ mapping. For every $t \in T$ consider the set

$$(6.4) \quad \mathcal{P}\left(t\right) = \left\{u \in C\left(S, X\right) : u\left(s\right) \in P\left(t, s\right) \ for \ every \ s \in S\right\}.$$

Notice that it is closed, convex and, by the Michael Selection Theorem, nonempty.

LEMMA 4. *The multifunction* $\mathcal{P} : T \longrightarrow clco\left\{C\left(S, X\right)\right\}$ *given by* (6.4) *is* $\mathcal{L} - measurable$.

PROOF. In view of the Proposition 36 it is enough to check that for any open ball $B = B\left(\varphi, r\right) \subset C\left(S, X\right)$ the set

$$\mathcal{P}^- B = \left\{t : \mathcal{P}\left(t\right) \cap B\left(\varphi, r\right) \neq \emptyset\right\} \in \mathcal{L}.$$

For this purpose notice that

$$(6.5) \qquad \mathcal{P}^- B = \bigcap_{s \in S}\left\{t : P\left(t, s\right) \cap B\left(\varphi\left(s\right), r\right) \neq \emptyset\right\}.$$

Indeed. If $t \in \mathcal{P}^- B$ then there exists $u \in \mathcal{P}\left(t\right) \cap B\left(\varphi, r\right)$. But this in particular means that for every $s \in S$ we have $u\left(s\right) \in P\left(t, s\right) \cap B\left(\varphi\left(s\right), r\right)$. Therefore

$$(6.6) \qquad \mathcal{P}^- B \subset \bigcap_{s \in S}\left\{t : P\left(t, s\right) \cap B\left(\varphi\left(s\right), r\right) \neq \emptyset\right\}.$$

On the other hand fix any $t \in \bigcap_{s \in S}\left\{t : P\left(t, s\right) \cap B\left(\varphi\left(s\right), r\right) \neq \emptyset\right\}$. Then, by the Theorem 43, the mapping $s \longrightarrow P\left(t, s\right)$ admits a continuous selection $u \in C\left(S, X\right)$ such that for every $s \in S$

$$u\left(s\right) \in P\left(t, s\right) \cap B\left(\varphi\left(s\right), r\right).$$

So
$$u \in \mathcal{P}(t) \cap B(\varphi, r)$$
and hence $t \in \mathcal{P}^- B$. But the latter means that

(6.7) $$\bigcap_{s \in S} \{t : P(t, s) \cap B[B(\varphi(s), r)] \neq \emptyset\} \subset \mathcal{P}^- B,$$

what together with (6.6) gives (6.5). Having this equality we get

(6.8) $$T \backslash \mathcal{P}^- B = \bigcup_{s \in S} \{t : P(t, s) \cap B[B(\varphi(s), r)] = \emptyset\} = proj_T A,$$

where $A = \{(t, s) : P(t, s) \cap B[B(\varphi(s), r)] = \emptyset\} \in \mathcal{L} \otimes \mathcal{B}$. So from the Theorem 6 we conclude that $\mathcal{P}^- B \in \mathcal{L}$. $\qquad \square$

Using the Lemma we shall prove

THEOREM 34. *Let S be a locally compact separable metric space. Then each l.C. multivalued mapping $P : T \times S \longrightarrow clco(X)$ admits a Carathéodory type selection.*

PROOF. If S is a compact metrizable space then $\mathcal{P}$ given by 6.4 is $\mathcal{L} - measurable$. Therefore it admits a measurable selection $\mathbf{p} : T \longrightarrow C(S, X)$. Hence $p(t, s) = \mathbf{p}(t)(s)$ is a required Carathéodory type selection.

In the general case we shall pick compact sets S_m, $m = 1, 2, \ldots$ such that
$$S_m \subset int S_{m+1} \text{ and } \bigcup_{m=1}^{\infty} S_m = S.$$

Then each P restricted to $T \times S_m$ admits a Carathéodory type selection p. We claim that p can be extended to a Carathéodory type selection on $T \times S_{m+1}$. For this purpose consider a multivalued mapping $P_m : T \times S_m \longrightarrow clco(X)$ given by

$$P_m(t, s) = \left\{ \begin{array}{ll} \{p(t, s)\} & for \quad (t, s) \in T \times S_m \\ P(t, s) & otherwise \end{array} \right\}$$

and observe that it is again l.C.. So it admits a Carathéodory type selection which is a required extension. Take a Carathéodory type selection on $T \times S_1$. Thus applying the above procedure we can extend it on the whole $T \times S$, what ends the proof. $\qquad \square$

CHAPTER 7

Fixed points property for convex-valued mappings

1. Single-valued case

The theory of fixed points for single-valued fuctions $p : X \longrightarrow X$ can also be developed for multivalued mappings of $P : X \longrightarrow N(X)$. We say that $x \in X$ is a fixed point of P iff

$$x \in P(x).$$

Such theory has been using in many areas, especially in nonlinear analysis, differential inclusions and control theory. There are many results concerning this subject and we do not pretend to give them all. We just present the most important and very useful ones. In this section we require convexity assumptions on the sets $P(x)$, nonconvex case can be discussed in the next part. We should say that in the fixed point theory for multifunctions a crucial role play selections. With the use of them many problems can be reduced to the single-valued case. Among all fixed point results for poinwise functions we shall mention the following:

THEOREM 35 (Schauder). *Let K be a compact convex subset of a Banach space X. Then each continuous mapping $f : K \longrightarrow K$ admits a fixed point.*

We should emphasize that despite convexity assumption the compactness is also essential. There are generalizations of the above result for arbitrary sets $K \in clco(X)$ but they require compactness of the function f. Recall that $f : X \longrightarrow X$ is said to be a compact mapping iff it is continuous and the set $f(X)$ is conditionally compact (totally bounded).

THEOREM 36. *Let K be a closed convex subset of a Banach space X. Then each compact mapping $f : K \longrightarrow K$ admits a fixed point.*

PROOF. Denote by $S = clco\{f(K)\}$ and observe that by the Mazur theorem S is compact. Fix $\varepsilon > 0$ and for every $s \in S$ consider sets

$$P_\varepsilon(s) = cl\{x \in K : |x - s| < d(s, K) + \varepsilon\}.$$

One can easily notice that they are nonempty and convex. Moreover, by the Example 6, the mapping

$$P_\varepsilon : S \longrightarrow clco\,(K)$$

is *l.s.c.* Choose a continuous selection $p_\varepsilon : S \longrightarrow K$ of P_ε and observe that for $s \in f\,(K)$ the inequality

(7.1) $$|p_\varepsilon\,(s) - s| \leq \varepsilon$$

holds. Therefore taking $f_\varepsilon = f \circ p_\varepsilon$ we have, for every $\varepsilon > 0$, a continuous function $f_\varepsilon : S \longrightarrow f\,(K) \subset S$ such that for each $s \in S$ is

$$|f_\varepsilon\,(s) - s| \leq \varepsilon.$$

Applying the Schauder Theorem we may, for every $\varepsilon > 0$, pick $s_\varepsilon \in f\,(K) \subset S$ such that

(7.2) $$f\,(p_\varepsilon\,(s_\varepsilon)) = s_\varepsilon.$$

But the net $\{s_\varepsilon\} \subset f\,(K) \subset S$ is totally bounded and we may assume that it is convergent, say

$$\lim_{\varepsilon \longrightarrow 0} s_\varepsilon = s_0.$$

Thus by, (7.2),

$$\lim_{\varepsilon \longrightarrow 0} p_\varepsilon\,(s_\varepsilon) = s_0$$

and taking the limits in (7.2) we conclude that s_0 is a fixed point of f. This completes the proof. $\qquad\square$

2. Multivalued case

The above two fixed point results have their counterparts for lower semicontinuous mappings. It is a simple application of the continuous selection method.

THEOREM 37. *Let Z be a compact convex subset of a Banach space X. Then each l.s.c. mapping $P : Z \longrightarrow clco\,(Z)$ admits a fixed point.*

PROOF. Take a continuous selection p of P. Then $p : Z \longrightarrow Z$ admits a fixed point x and it is also a fixed point of P since $x = p\,(x) \in P\,(x)$. $\qquad\square$

Using exactly the same arguments we get the following generalization of the Theorem 1.9:

THEOREM 38. *Let Z be a closed convex subset of a Banach space X. Then each l.s.c. mapping $P : Z \longrightarrow clco\,(Z)$ such that $P(Z)$ is conditionally compact admits a fixed point.*

The continuous selection method can also be used for upper semicontinuous mappings but we can not do that directly since upper semicontinuity do not guarrantee the existence of continuous selections. Provided below proof is based on the following observation in a metrizable space S.

LEMMA 5. *For given* $P : S \longrightarrow N(X)$, $s \in S$ *and* $r > 0$ *set*

$$O_r(s) = \bigcup_{z \in B(s,r)} P(z) = P\{B(s,r)\}.$$

Then the mapping $O_r : S \longrightarrow N(X)$ *is l.s.c.*

PROOF. We shall show that for any s_0, $x_0 \in O_r(t_0)$ and a net $s_\alpha \longrightarrow s_0$ one can find a net $x_\alpha \in O_r(s_\alpha)$ such that $x_\alpha \longrightarrow x_0$. To do that pick $z_0 \in B(s_0, r)$ such that $x_0 \in P(z_0)$. Since $s_\alpha \longrightarrow s_0$ then for "sufficiently large" α we have $z_0 \in B(s_\alpha, r)$. Therefore $x_0 \in \bigcup_{z \in B(s_\alpha, r)} P(z) = O_r(s_\alpha)$ and we can simply take $x_\alpha = x_0$ for "sufficiently large" α. $\square$

Having the Lemma we can present a promised fixed point result for upper semicontinuous mapping.

THEOREM 39 (Ky Fan). *Let* Z *be a compact convex subset of a Banach space* X. *Then each u.s.c. mapping* $P : Z \longrightarrow clco(Z)$ *admits a fixed point.*

PROOF. The theorem 38 applied to each

$$O_n(x) = clcoP\left\{B\left(x, \frac{1}{n}\right)\right\}$$

yields the existence of a sequence $\{x_n\}_{n \in N} \subset Z$ such that

$$x_n \in clcoP\left\{B\left(x_n, \frac{1}{n}\right)\right\}.$$

Since Z is compact we may assume that $x_n \longrightarrow x_0$. We shall show that

$$x_0 \in P(x_0).$$

For doing this it is sufficient to verify that for any $x^* \in B^*$ the inequality

$$\langle x^*, x_0 \rangle \leq c_P(x_0, x^*)$$

holds. For this purpose we calculate

$$\langle x^*, x_n \rangle \leq c_{clcoP\left\{B\left(x_n, \frac{1}{n}\right)\right\}}(x^*) = c_{P\left\{B\left(x_n, \frac{1}{n}\right)\right\}}(x^*) =$$

$$= \sup\left\{ \langle x^*, u \rangle : u \in P\left\{ B\left(x_n, \frac{1}{n} \right) \right\} \right\} \leq \sup_{z \in B\left(x_n, \frac{1}{n} \right)} c_P\left(z, x^* \right).$$

Therefore for each n we may choose $z_n \in B\left(x_n, \frac{1}{n} \right)$ such that

$$\langle x^*, x_n \rangle - \frac{1}{n} < c_P\left(z_n, x^* \right).$$

But then $z_n \longrightarrow x_0$ and hence

$$\langle x^*, x_0 \rangle \leq \limsup c_P\left(z_n, x^* \right) \leq c_P\left(x_0, x^* \right),$$

what has to be shown.   □

3. Parametrized fixed point results

In this section we shall deal with mappings $P : T \times Z \longrightarrow clco\left(Z \right)$ defined on a compact convex subset Z of a separable Banach space X. We always postulate separability of X. However, in the general situation we can restrict ourselves to $X_0 = spanZ \subset X$ being a separable Banach space and providing all considerations for X_0.

Assuming that for each $t \in T$ mapping $x \longrightarrow P\left(t, x \right)$ admits the fixed points we may consider a nonempty set

(7.3) $$\mathcal{F}\left(t \right) = \{ x : x \in P\left(t, x \right) \}$$

and examine next some properties of a new multifunction $\mathcal{F} : T \longrightarrow N\left(Z \right)$. Especially we are interested in the measurability of $\mathcal{F}$ and the existence of measurable selections. This time we begin with $u.C.$ mappings and next apply them to derive some properties in $l.C.$ case.

PROPOSITION 47. *Take any upper Carathéodory type mapping $P : T \times Z \longrightarrow clco\left(KZ \right)$ and consider sets $\mathcal{F}\left(t \right)$ given by (7.3). Then each $\mathcal{F}\left(t \right)$ is compact and $\mathcal{F} : T \longrightarrow c\left(Z \right)$ is a measurable multifunction.*

PROOF. Since Z is compact then for every $t \in T$ the set

$$grP\left(t, \cdot \right) = \{ (x, z) \in Z \times Z : z \in P\left(t, x \right) \}$$

is compact and this implies the compactness of each set

$$\mathcal{F}\left(t \right) = proj_X \left\{ grP\left(t, \cdot \right) \right\}.$$

To see the measurability we shall use twice the Scorza - Dragoni type Theorem 31. For every $\varepsilon > 0$ there exists a compact set $T_\varepsilon \subset T$ with $\mu\left(T \backslash T_\varepsilon \right) < \varepsilon$ such that $P : T_\varepsilon \times Z \longrightarrow clco\left(K \right)$ is $u.s.c.$. Hence it's graph restricted to $T_\varepsilon \times Z$,

$$\mathcal{G} = \{ (t, x, z) : (t, x) \in T_\varepsilon \times Z \text{ and } z \in P\left(t, x \right) \},$$

is compact and therefore closed. Thus $\mathcal{F}$ restricted to T_ε has the closed graph since

$$gr\mathcal{F} = \{(t, x, x) : (t, x) \in T_\varepsilon \times Z \text{ and } x \in \mathcal{F}(t)\} =$$
$$= \{(t, x, z) \in \mathcal{G} : z = x\}.$$

But this means that $\mathcal{F}$ restricted to each T_ε is $u.s.c.$, what yields the measurability of $\mathcal{F}$ on T. $\qquad\qquad\qquad\qquad\qquad\qquad\qquad\square$

COROLLARY 7. *Each $u.C.$ type mapping $P : T \times Z \longrightarrow clco(Z)$ admits a measurable mapping $x : T \longrightarrow X$ such that for all $t \in T$*

$$x(t) \in P(t, x(t)).$$

In particular, for each Carathéodory type function $p : T \times Z \longrightarrow clco(Z)$ there is a measurable mapping $x : T \longrightarrow X$ such that for all $t \in T$

$$x(t) = p(t, x(t)).$$

Measurable selections of fixed points exist also for $l.C.$ type mappings. Since any $l.C.$ type mapping $P : T \times Z \longrightarrow clco(K)$ admits a Carathéodory type selection then the previous result implies

COROLLARY 8. *Each $l.C.$ type mapping $P : T \times Z \longrightarrow clco(Z)$ admits a measurable mapping $x : T \longrightarrow X$ such that for all $t \in T$*

$$x(t) \in P(t, x(t)).$$

Part 3

DECOMPOSABILITY

CHAPTER 8

Decomposable sets

1. General properties

In this part we restrict our considerations to a Banach space X with separable X^* and $(T, \mathcal{L}, \mu)$ a measure space defined on a complete separable metric space with a $\sigma - field$ L of Lebesgue measurable sets given by a finite Radon measure μ. Without loss of generality we may assume that μ is probabilistic, i.e. $\mu(T) = 1$.

Let $P : T \longrightarrow N(X)$ be a multifunction and consider a set

$$K_P = \{u \in M(T, X) : u(t) \in P(t) \quad a.e. \ in \ T\}$$

or

$$K_P = \{u \in L^p(T, X) : u(t) \in P(t) \quad a.e. \ in \ T\}.$$

If we assume that $P : T \longrightarrow co(X)$, then the set K_P is convex as well. This property can not be preseved for non-convex valued multifunctions $P : T \longrightarrow N(X)$. But in any case the set K_P satisfies the following property:

for all $u, v \in K_P$ *and any* $A \in \mathcal{L}$ *we have*

(Q) $$\chi_A u + (1 - \chi_A) v \in K_P.$$

DEFINITION 18. *A subset K of $M(T, X)$ or $L^p(T, X)$ satisfying (Q) is said to be decomposable. The family of all nonempty decomposable and closed decomposable subsets of $M(T, X)$ $\{L^p(T, X)\}$ we shall denote, respectively, by $dec_0(T, X)$ $\{dec_p(T, X)\}$ and $dcl_p(T, X)$ $\{dcl_p(T, X)\}$. If $p = 1$ then we write $dec(T, X)$ and $dcl(T, X)$, respectively.*

The notion of decomposability was introduced by Rockafellar [205] in 1968 in connections with integral functionals and since than decomposable sets became a main tool in nonconvex analysis. They are in a sense a subsitute of convexity and many properties of convex sets have counterparts for decomposable sets. The families $dec_p(T, X)$ and $dcl_p(T, X)$ are closed with respect to intersection and closure. Namely, we have:

115

PROPOSITION 48. *Decomposable sets possess the following properties:*

(i)   *Let* $\{K_\alpha\}_{\alpha \in \Lambda} \subset dec_p(T, X)$. *Then* $\bigcap_{\alpha \in \Lambda} K_\alpha \in dec_p(T, X)$.
Similar property holds for $dcl_p(T, X)$.

(ii)   *Let* $\{K_n\}_{n \in N} \subset dec_p(T, X)$ *is an increasing sequence. Then* $\bigcup_{n=1}^{\infty} K_n \in dec_p(T, X)$.

(iii)   *If* $K \in dec_p(T, X)$, *then* $clK \in dcl_p(T, X)$.

(iv)   *For* $K \in dec_p(T, X)$ *and any* $u \in L^p(T, X)$ *the set* $K - u \in dec_p(T, X)$. *Similarly, for* $K \in dcl_p(T, X)$ *we have* $K - u \in dcl_p(T, X)$.

PROOF. (i)   Take $u, v \in \bigcap_{\alpha \in \Lambda} K_\alpha$ and $A \in \mathcal{L}$. Then for every $\alpha \in \Lambda$ we have $\chi_A u + (1 - \chi_A) v \in K_\alpha$ and therefore

$$\chi_A u + (1 - \chi_A) v \in \bigcap_{\alpha \in \Lambda} K_\alpha.$$

(ii)   Take $u, v \in \bigcup_{n=1}^{\infty} K_n$ and $A \in \mathcal{L}$. Since the collection $\{K_n\}_{n \in N}$ is increasing therefore there exists n such that both $u, v \in K_n$. Hence $\chi_A u + (1 - \chi_A) v \in K_n \subset \bigcup_{n=1}^{\infty} K_n$.

(iii)   Take $u, v \in clK$ and $A \in \mathcal{L}$. There exist $u_n \in K$ and $v_n \in K$ such that $u_n \longrightarrow u$ and $v_n \longrightarrow v$. But $\chi_A u_n + (1 - \chi_A) v_n \longrightarrow \chi_A u + (1 - \chi_A) v$ and $\chi_A u_n + (1 - \chi_A) v_n \in K$, so $\chi_A u + (1 - \chi_A) v \in clK$.

(iv)   It is straighforward.   □

Since $M(T, X)$ and $L^p(T, X)$ are decomposable so the family of decomposable sets containing given $K \in dec_p(T, X)$ is nonempty. Therefore from the Proposition 48 follows that for any $K \subset L^p(T, X)$ there exist the smallest decomposable and closed decomposable set containing K. We shall call them, respectively, decomposable hull and closed decomposable hull and denote by $dec_p(K)$ and $dcl_p(K)$. Obviously we have

$$K \subset dec_p(K) \subset dcl_p(K).$$

PROPOSITION 49. *Let* $u, v \in L^p$ *be given. Then*

$$dec_p\{u, v\} = dcl_p\{u, v\} = \{\chi_A u + (1 - \chi_A) v : A \in \mathcal{L}\}.$$

PROOF. Since $\{u\chi_A + (1 - \chi_A) v : A \in \mathcal{L}\}$ is closed set containing u and v then

$$dec_p\{u, v\} \subset dcl_p\{u, v\} \subset \{\chi_A u + (1 - \chi_A) v : A \in \mathcal{L}\}.$$

The opposite inclusions follows from the observation that for arbitrary $A, B, C \in \mathcal{L}$ we have

$$\chi_C \{\chi_A u + (1 - \chi_A) v\} + (1 - \chi_C) \{\chi_B u + (1 - \chi_B) v\} =$$
$$= \chi_D u + (1 - \chi_D) v,$$

where $D = A \cap C \cup B \cap (T \backslash C)$.   □

PROPOSITION 50. *Let* $K \in dec_p(T, X)$, $1 \le p < \infty$, *be given. Then*

a) *for every finite* $\{u_k\}_{k=1}^n \subset K$ *and each finite partition* $\{A_k\}_{k=1}^n$ *of* T *the function* $\sum\limits_{k=1}^n \chi_{A_k} u_k \in K$;

b) *if* $K \subset L^p(T, X)$ *is* p − *integrably bounded then also for every* $\{u_k\}_{k=1}^\infty \subset K$ *and each partition* $\{A_k\}_{k=1}^\infty$ *of* T *we have*

$$\sum_{k=1}^\infty \chi_{A_k} u_k \in dcl_p(K).$$

PROOF. We shall apply an induction argument. For $n = 2$ it simply follows from the definition.

Assume that for some $n \ge 2$ the condition a) holds. Take arbitrary $\{u_k\}_{k=1}^{n+1} \subset K$ and any partition $\{A_k\}_{k=1}^{n+1}$ of T. Denote by

$$B_i = A_i, \ i = 1, ..., n - 1 \ and \ B_n = A_n \cup A_{n+1}$$

and observe that $\{B_k\}_{k=1}^n$ is a partition of T. Therefore

$$w = \sum_{k=1}^n \chi_{B_k} u_k \in K.$$

But $w = \sum\limits_{k=1}^n \chi_{A_k} u_k + \chi_{A_{n+1}} u_n$ and hence $\left(1 - \chi_{A_{n+1}}\right) w = \sum\limits_{k=1}^n \chi_{A_k} u_k$. Therefore

$$\sum_{k=1}^{n+1} \chi_{A_k} u_k = \chi_{A_{n+1}} u_{n+1} + \left(1 - \chi_{A_{n+1}}\right) w \in K,$$

what gives an induction step for $n + 1$.

To show b) let the set $K \subset L^p(T, X)$ be p−integrably bounded by $\varphi \in L^p(T, R)$. By the Proposition 48iii. we may assume that

$$0 \in K.$$

Fix $\{u_k\}_{k=1}^\infty \subset K$ and a partition $\{A_k\}_{k=1}^\infty$ of T. We need to show that $\sum\limits_{k=1}^\infty \chi_{A_k} u_k \in dcl_p(K)$. Take any $\varepsilon > 0$ and let n be large enough such that we have

$$\sum_{k=n+1}^{\infty} \int_{A_k} a^p(t)\,\mu(dt) = \int_{\substack{\infty \\ \bigcup_{k=n+1} A_k}} |\varphi(t)|^p\,\mu(dt) < \varepsilon^p.$$

The existence of such n follows from the Vitaly-Hahn-Saks Theorem, since

$$\lim_{n \longrightarrow \infty} \mu \left\{ \bigcup_{k=n+1}^{\infty} A_k \right\} = 0.$$

Now

$$v_n = \sum_{k=1}^{n} \chi_{A_k} u_k = \sum_{k=1}^{n} \chi_{A_k} u_k + \chi_{A_{k+1}} \cdot 0 \in K$$

and we may proceed as follows:

$$\left\| \sum_{k=1}^{\infty} \chi_{A_k} u_k - v_n \right\|_p^p = \left\| \sum_{k=n+1}^{\infty} \chi_{A_k} u_k \right\|_p^p \leq \sum_{k=n+1}^{\infty} \int_{A_k} |\varphi(t)|^p\,\mu(dt) < \varepsilon^p.$$

So

$$d_p \left(\sum_{k=1}^{\infty} \chi_{A_k} u_k, K \right) < \varepsilon.$$

Since $\varepsilon > 0$ was arbitrarily chosen, we conclude that $\sum_{k=1}^{\infty} \chi_{A_k} u_k \in dcl_p(K).$ $\qquad\square$

As we have already mentioned decomposable sets have many nice properties. However sometimes they also may behave in an unpleasent way. Any decomposable set smaller than $L^p(T, X)$ has no interior points and any compact decomposable set reduces to a point.

PROPOSITION 51. *Let B be the unit ball in $L^p(T, X)$, where $1 \leq p < \infty$. Then $dec_p(B) = L^p(T, X)$.*

PROOF. Take arbitrary $u \in L^p(T, X)$. We claim that $u \in dec_p(B)$. Indeed. Fix an integer n such that $\|u\|_p^p \leq n$. Let $\{A_r\}_{r \in I}$ be a segment for a real measure $m(A) = \int_A |u(t)|^p\,\mu(dt)$. For $k = 1, ..., n$ denote by $B_k = A_{\frac{k}{n}} \backslash A_{\frac{k-1}{n}}$ and observe that $\{B_k\}_{k=1}^n$ is a partition of T with $m(B_k) = \frac{1}{n} \int_T |u(t)|^p\,\mu(dt) \leq 1$. But then functions $u_k = \chi_{B_k} u \in B$

and hence our claim follows from an identity

$$\sum_{k=1}^{n} \chi_{B_k} u_k = \sum_{k=1}^{n} \chi_{B_k} u = u.$$

$\square$

PROPOSITION 52. *Let $K \in dec_p(T,X)$ be a relatively compact subset of $L^p(T,X)$. Then K reduces to one point.*

PROOF. By the Proposition 48*iii*. we may require that $0 \in K$. To a contrary assume that there is in K a point $u \neq 0$. One can therefore find $\varepsilon > 0$, small enough, such that the set

$$B = \{t : |u(t)| \geq \varepsilon\}$$

is of positive measure. Notice that $u\chi_B \in K$ and thus

$$K_0 = \{u\chi_A : A \in \mathcal{L}_{|B}\} = \{u\chi_{A \cap B} : A \in \mathcal{L}\} = dcl_p\{u\chi_B, 0\} \subset clK$$

is a compact subset of $L^p(T,X)$. Now observe that for $A_1, A_2 \in \mathcal{L}_{|B}$ we have

$$\|A_1 - A_2\|_1 \leq \frac{1}{\varepsilon^p} \|u\chi_{A_1} - u\chi_{A_2}\|_p^p$$

and hence the mapping $u\chi_A \longrightarrow A$ is well-defined and continuous from K_0 onto $\mathcal{L}_{|B}$. But this implies that $\mathcal{L}_{|B}$ is compact, what in turn gives a contradiction. $\square$

There is a strong relation between closed decomposable subsets $K \subset L^1(T,X)$ and closed-valued measurable multivalued mappings. Namely we have:

THEOREM 40. *A closed subset $K \in dcl(T,X)$ is decomposable iff there is a measurable $P : T \longrightarrow cl(X)$ such that $K = K_P$.*

PROOF. $\Longleftarrow$
Since the decomposability of K_P is obvious, we need only to show that it is closed. But it can be easily deduced from the fact that any convergent in $L^1(T,X)$ sequence contains a subsequence convergent almost everywhere.

$\Longrightarrow$
Since $K \subset L^1(T,X)$ is closed therefore it is also separable. Thus K admits a dense subset $\{p_n\}_{n=1}^{\infty}$. Take

$$P(t) = cl\{p_n(t) : n = 1, 2, ...\}$$

and observe that $P : T \longrightarrow cl(X)$ is measurable. One can easily check that $K_P \subset K$. We shall show that the opposite inclusion holds. To see this it is enough to check that for every $\varepsilon > 0$ and $u \in K$ we have $K_P \cap B(u, \varepsilon) \neq \emptyset$. For this purpose fix $u \in K$ and $\varepsilon > 0$. By the density of $\{p_n\}_{n=1}^\infty$ there exists a subsequence $\{p_{k_n}\}_{n=1}^\infty$ converging to u. Passing, if necessary, again to a subsequence we may assume that $\{p_{k_n}\}_{n=1}^\infty$ also tends to u a.e. in T. The latter in particular means that

$$T_0 = \left\{ t : P(t) \cap B\left\{ u(t), \frac{\varepsilon}{2} \right\} \neq \emptyset \right\} =$$

$$(8.1) \qquad = \left\{ t : d(u(t), P(t)) < \frac{\varepsilon}{2} \right\} = \bigcup_{n=1}^\infty \left\{ t : |u(t) - p_n(t)| < \frac{\varepsilon}{2} \right\}$$

is a set of full measure. Denote by

$$T_n = \left\{ t : |u(t) - p_n(t)| < \frac{\varepsilon}{2} \right\}$$

and observe that T_n are such measurable sets that

$$\bigcup_{n=1}^\infty T_n = T_0.$$

Consider the sets A_n given by:

$$A_0 = T \backslash T_0, \ A_1 = T_1, \ A_n = T_n \backslash \bigcup_{k=1}^{n-1} A_k \ \ for \ \ n \geq 2.$$

Then $\{A_n\}_{n=0}^\infty$ is a partition of T and therefore

$$(8.2) \qquad \mu\left(T \backslash \bigcup_{n=1}^k A_n \right) \xrightarrow[k \to \infty]{} 0.$$

Moreover, for $t \in A_n$, we have

$$(8.3) \qquad |u(t) - p_n(t)| < \frac{\varepsilon}{2}.$$

By the Vitaly-Hahn-Saks Theorem (see [62]) and (8.2) one can choose a positive integer k such that

$$(8.4) \qquad \| p_1 \chi_{T \backslash \bigcup_{n=1}^k A_n} \|_1 = \int_{T \backslash \bigcup_{n=1}^k A_n} |p_1(t)| \, \mu(dt) < \frac{\varepsilon}{2}.$$

Consider a function p given by

$$p = \sum_{n=1}^k p_n \chi_{A_n} + p_1 \chi_{T \backslash \bigcup_{n=1}^k A_n}.$$

Obviously $p \in K_P$. We claim that $p \in B(u, \varepsilon)$. To see this observe that, by (8.3) and (8.4), we have

$$\| u - p \|_1 = \sum_{n=1}^{k} \int_{A_n} |u(t) - p_n(t)| \, \mu(dt) + \| p_1 \chi_{T \setminus \bigcup_{n=1}^{k} A_n} \|_1 < \varepsilon,$$

what completes the proof. $\square$

COROLLARY 9. *Let* $K = K_P \in dec(T, X)$ *be a decomposable set given by an* $\mathcal{L} - measurable$ *multifunction* $P : T \longrightarrow N(X)$. *Then* $clK = K_{clP}$, *where* clP *is defined by* $(clP)(t) = cl(P(t))$.

PROOF. The Theorem 40 yields the existence of a measurable $P_0 : T \longrightarrow cl(X)$ such that $clK = K_{P_0}$. Clearly $P(t) \subset P_0(t)$ *a.e. in* T and hence

(8.5) $\qquad (clP)(t) = cl(P(t)) \subset P_0(t)$ *a.e. in* T.

On the other hand K_{clP} is closed decomposable set containing K and therefore

$$K_{P_0} = clK \subset K_{clP},$$

what together with (8.5) means that $P_0(t) = (clP)(t)$ *a.e. in* T. $\square$

2. Decomposable sets in $L^p(T, R)$

A special role play decomposable subsets of $L^p(T, R)$. The examination of their topological properties we begin with the following:

PROPOSITION 53. *Let* $K \in dec_p(T, R)$. *Then for every* $\{u_i\}_{i=1}^{n} \subset K$ *the functions*

$$\min_{1 \le i \le n} u_i \in K \quad and \quad \max_{1 \le i \le n} u_i \in K.$$

If, additionally, K *is p-integrably bounded from below then for every* $\{u_i\}_{i=1}^{\infty} \subset K$ *the functions*

$$\inf_{n} u_n \in cl(K).$$

Similarly, if K *is p-integrably bounded from above then*

$$\sup_{n} u_n \in cl(K).$$

PROOF. To show the first part it is enough to consider the case $n = 2$ and next apply an induction argument. Denote by

$$A = \{t : u_1(t) < u_2(t)\}.$$

Now our claim follows from an easy observation that

$$\min (u_1, u_2)(t) = \begin{cases} u_1(t) & if \ t \in A \\ u_2(t) & if \ t \in T\backslash A \end{cases} =$$

$$= \chi_A u_1 + (1 - \chi_A) u_2$$

and

$$\max (u_1, u_2)(t) = \begin{cases} u_2(t) & if \ t \in A \\ u_1(t) & if \ t \in T\backslash A \end{cases} =$$

$$= \chi_A u_2 + (1 - \chi_A) u_1.$$

Assume now that K is p-integrably bounded from below by $a \in L^p(T, R)$ and let $\{u_i\}_{i=1}^n \subset K$ be given. Therefore for each n the functions

$$a_n = \min_{1 \le i \le n} u_i \in K$$

and the inequalities

$$a_1(t) \ge ... \ge a_n(t) \ge ... \ge a(t) \quad a.e. \ in \ T$$

hold. The latter, in view of the Lebesgue Dominated Convergence Theorem, means that $\{a_n\}_{n=1}^\infty$ is convergent in L^p to $\inf a_n = \inf u_n$. Thus $\inf_n u_n \in cl(K)$. The arguments for the third part are analogous as for the second. This ends the proof. $\qquad\square$

As a conclusion from previous considerations and the Proposition 12 we have

COROLLARY 10. *Let* $K \in dec_p(T, R)$ *be* $p - integrably \ bounded$ *from below. Then*

$$ess \inf \{u : u \in K\} \in cl(K).$$

Similarly, if K *is p-integrably bounded from above then*

$$ess \sup \{u : u \in K\} \in cl(K).$$

For $K \in dec_p(T, X)$ denote by $|K| = \{|u(\cdot)| : u \in K\}$ and observe that $|K| \in dec_p(T, R)$. Since $|K|$ is clearly p-integrably bounded from below then the Corollary 10 yields
(8.6)
$$\varphi(t) = ess \inf \{|u(t)| : u \in K\} = ess \inf \{w(t) : w \in |K|\} \in cl(|K|).$$

Obviously $\varphi \in L^p(T, R)$. The function φ plays the role of pointwise distance of K from 0. This is explained in the following:

PROPOSITION 54. Let $K \in dcl_p(T, X)$ and consider $v \in L^p(T, R)$ satisfying the inequality

$$\varphi(t) < v(t) \quad a.e. \ in \ T.$$

Then there exists $u \in K$ such that

(8.7) $$|u(t)| < v(t) \quad a.e. \ in \ T.$$

Moreover, the relation

(8.8) $$\|\varphi\|_p = d_p(0, K)$$

holds.

PROOF. Applying the Theorem 12 we have the existence of $u_n \in K$ with

(8.9) $$|u_1(t)| \geq |u_2(t)| \geq \ldots \geq |u_n(t)| \geq \ldots \geq \varphi(t) \quad a.e. \ in \ T$$

and

(8.10) $$\lim_{n \longrightarrow \infty} |u_n(t)| = \varphi(t) \quad a.e. \ in \ T.$$

Consider measurable sets

$$T_n = \{t : |u_n(t)| < v(t)\}$$

and notice that by (8.9) and (8.10) the family $\{T_n\}$ is increasing and

$$\bigcup_{n=1}^{\infty} T_n = T_0$$

is a set of full measure. Then the family $\{A_n\}_{n=0}^{\infty}$ given by

$$A_0 = T \backslash T_0, \ A_1 = T_1, \ A_n = T_n \backslash \bigcup_{k=1}^{n-1} A_k \quad for \ n \geq 2$$

forms a partition of T. Notice that by (8.9) the set

$$K_0 = dec_p\{u_n : n = 1, 2, \ldots\}$$

is $p-$integrably bounded and

$$K_0 \subset dcl_p\{u_n : n = 1, 2, \ldots\} \subset K.$$

Thus, by the Proposition 50,

$$u = \sum_{n=1}^{\infty} \chi_{A_n} u_n \in K.$$

One can easily check that (8.7) holds.

We shall now demonstrate that (8.8) is satisfied. First of all let us notice that

$$\|\varphi\|_p \leq d_p(0, K).$$

To see the opposite inequality take arbitrary $\varepsilon > 0$ and let $u \in K$ be such that

$$|u(t)| < |\varphi(t)| + \varepsilon \quad a.e. \ in \ T.$$

Thus

$$d_p(0, K) \leq \|u\|_p \leq \|\varphi\|_p + \varepsilon.$$

Hence passing to the limit with $\varepsilon \longrightarrow 0$ one gets

$$d_p(0, K) \leq \|\varphi\|_p,$$

what ends the proof. $\square$

Let $K \in dcl_p(T, X)$. For every $w \in L^p(T, X)$ put

(8.11) $\varphi(w)(t) = ess \inf \{|u(t) - w(t)| : u \in K\}.$

Since $\varphi(w)(t) = ess \inf \{|u(t)| : u \in K - w\}$ and $K - w \in dcl_p(T, X)$ then, by the Proposition 53,

$$\|\varphi(w)\|_p = d(w, K).$$

So (8.11) defines a mapping $\varphi : L^p(T, X) \longrightarrow L^p(T, R)$.

PROPOSITION 55. *The mapping* $\varphi : L^p(T, X) \longrightarrow L^p(T, R)$ *given by (8.11) is Lipschitz with constant 1.*

PROOF. Take any $u, v \in L^p(T, X)$ and fix $\varepsilon > 0$. By the previous Proposition 54 there are $w, z \in K$ such that

$$|u(t) - w(t)| < \varphi(u)(t) + \varepsilon \quad a.e. \ in \ T$$

and

$$|v(t) - z(t)| < \varphi(v)(t) + \varepsilon \quad a.e. \ in \ T.$$

Moreover, by the definition of φ, we have

$$\varphi(u)(t) \leq |u(t) - z(t)| \quad a.e. \ in \ T$$

and

$$\varphi(v)(t) \leq |v(t) - w(t)| \quad a.e. \ in \ T.$$

Then

$$\varphi(v)(t) - \varphi(u)(t) \leq |v(t) - w(t)| - |u(t) - w(t)| + \varepsilon \leq$$
$$\leq |v(t) - u(t)| + \varepsilon \quad a.e. \ in \ T.$$

Similarly

$$\varphi(u)(t) - \varphi(v)(t) \leq |u(t) - z(t)| - |v(t) - z(t)| + \varepsilon \leq$$
$$\leq |u(t) - u(t)| + \varepsilon \quad a.e. \ in \ T$$

and therefore
$$\|\varphi(u) - \varphi(v)\|_p \leq \|u - v\|_p + \varepsilon.$$
But $\varepsilon > 0$ is arbitrary, so the passing to the limit with $\varepsilon \longrightarrow 0$ we obtain our claim.   □

At the end we have to say that in many situations the examination of decomposable subsets of $L^p(T, X)$ can be reduced to $L^1(T, X)$ via the Mazur transformation given (4.4). This is explained in the following:

PROPOSITION 56. *A set $K \in dec_p(T, X)$ iff $\omega_p(K) \in dec(T, X)$. Similarly $K \in dcl_p(T, X)$ iff $\omega_p(K) \in dcl(T, X)$.*

We leave the proofs for the reader.

3. A separation result

For decomposable sets a kind of separation result holds. Making analogous consideration as in linear case one can prove the following:

THEOREM 41. *Consider two disjoint decomposable sets $K_1, K_2 \in dec(T, X)$. Then $L^1(T, X)$ can be split into two disjoint decomposable subsets $L_1, L_2 \in dec(T, X)$ such that $K_i \subset L_i$, $i = 1, 2$.*

PROOF. Consider a family $\mathcal{K} = \{L \in dec(T, X) : L \cap K_2 = \emptyset\}$. The family $\mathcal{K}$ is nonempty, since $K_1 \in \mathcal{K}$. Order the family $\mathcal{K}$ by inclusion and observe that any increasing family $\{L_n\} \subset \mathcal{K}$ has an upper bound. Indeed, this rule plays $\bigcup_{n=1}^{\infty} L_n$, which obviously is decomposable. By the Kuratowski-Zorn Lemma in $\mathcal{K}$ there is an maximal element $L_1 \in \mathcal{K}$ and by construction
$$K_1 \subset L_1.$$
Take $L_2 = L^1(T, X) \setminus L_1$ and observe that
$$K_2 \subset L_2.$$
We need to show that L_2 is decomposable. If it is not the case, then there exist $u, v \in L_2$ and $A \in \mathcal{L}$ such that
$$w = \chi_A u + (1 - \chi_A)v \notin L_2.$$
Then $w \notin K_2$ and hence at least one of u, v is not a member of K_2. Indeed, if both $u, v \in K_2$ then by decomposability $w = \chi_A u + (1 - \chi_A)v \in K_2$, a contradiction. So, assume for example that $u \notin K_2$. Thus $dec\{u, L_1\} \in \mathcal{K}$ and by maximality $u \in L_1$, what contradicts with the choice of u. So, the proof is complete.   □

CHAPTER 9

Selections

1. Continuous selections

We will discuss the continuous selection property for a lower semi-continuous multifunction with closed, decomposable values (Bressan-Colombo-Fryszkowski Theorem) and their consequences.

We shall be considering $l.s.c.$ mappings $P : S \longrightarrow dec\,(T, X)$ defined on **a separable metric space** S and with decomposable values. We begin with some properties of a mapping $\psi : S \longrightarrow L^1\,(T, R)$ given by

$$\psi\,(s) = ess\inf\left\{|u\,(t)| : u \in P\,(s)\right\}.$$

In virtue of the Proposition 54 for every $v \in L^1\,(T, R)$ such that

$$v\,(t) > \psi\,(s)\,(t) \quad a.e. \ in \ T$$

there exists $u \in P\,(s)$ such that

$$|u\,(t)| > \psi\,(s)\,(t) \quad a.e. \ in \ T.$$

Therefore for every $s \in S$ the set

$$(9.1) \qquad R\,(s) = \left\{v \in L^1\,(T, R) : v\,(t) > \psi\,(s)\,(t) \quad a.e. \ in \ T\right\} \neq \emptyset$$

and one can easily check that it is decomposable and convex.

PROPOSITION 57. *Let* $P : S \longrightarrow dec\,(T, X)$ *be a l.s.c. multifunction. Then the mapping* $R : S \longrightarrow dec\,(T, R) \cap co\,(L^1\,(T, R))$, *given by (9.1), is l.s.c..*

PROOF. Let $F \subset L^1\,(T, R)$ be an arbitrary closed subset. We need to show that $R^+ F$ is closed. For this purpose take any sequence $s_n \longrightarrow s_0$ such that

$$R\,(s_n) \subset F \quad for \quad n = 1, 2, \dots$$

and choose any $v_0 \in R\,(s_0)$. One can pick $u_0 \in P\,(s_0)$ such that

$$|u_0\,(t)| > \psi\,(s_0)\,(t) \quad a.e. \ in \ T.$$

By the $l.s.c.$ of P there are $u_n \in P\,(s_n)$ such that $u_n \longrightarrow u_0$. Consider the functions $v_n = |u_n| + v_0 - |u_0| + \frac{1}{n}$. Notice that $v_n \in R\,(s_n) \subset F$ and $v_n \longrightarrow v_0$ in $L^1\,(T, R)$. Therefore $v_0 \in F$, what completes the proof. $\qquad \square$

127

PROPOSITION 58. *Let $P : S \longrightarrow dec\,(T, X)$ be a l.s.c. multifunction and assume that there exist continuous mappings $\varphi : S \longrightarrow L^1\,(T, X)$ and $r : S \longrightarrow L^1\,(T, R)$ such that for every $s \in S$ the set*

$$G(s) = \{u \in P(s) : |u(t) - \varphi(s)(t)| < r(s)(t) \quad a.e.\ in\ T\} \neq \emptyset.$$

Then $G : S \longrightarrow dec\,(T, X)$ is a l.s.c. multifunction.

PROOF. Obviously each $G(s)$ is decomposable. Since $P(s) - \varphi(s)$ is *l.s.c.* then shifting P if neccesary we may assume that $\varphi \equiv 0$. So we may assume that

$$G(s) = \{u \in P(s) : |u(t)| < r(s)(t) \quad a.e.\ in\ T\}.$$

In order to show that G is *l.s.c.* take arbitrarily a closed set $F \subset L^1\,(T, R)$. We need to verify that G^+F is closed. For this purpose choose any sequence $s_n \longrightarrow s_0$ such that:

$$G(s_n) \subset F \quad for\ \ n = 1, 2, \ldots$$

and pick up any $u_0 \in G(s_0)$. Therefore

$$|u_0(t)| < r(s_0)(t) \quad a.e.\ in\ T.$$

By the *l.s.c.* of P there are $u_n \in P(s_n)$, $n = 1, 2, \ldots$, such that $u_n \longrightarrow u_0$. By taking if necessary subsequences we may assume that

$$u_n \longrightarrow u_0 \quad and \quad r(s_n) \longrightarrow r(s_0) \quad a.e.\ in\ T.$$

Applying the Egorov Theorem we have the existence of an increasing sequence $\{T_i\}$ of measurable sets that

$$\int_{T \setminus T_i} r(s_0)(t)\,\mu(dt) < \frac{1}{i}$$

and

$$u_n \rightrightarrows u_0 \quad and \quad r(s_n) \rightrightarrows r(s_0) \quad uniformly\ on\ T_i.$$

Consider the sets

$$T_{i,k} = \left\{ t \in T_i : |u_0(t)| < r(s_0)(t) - \frac{1}{k} \quad a.e.\ in\ T \right\}$$

and notice that $\{T_{i,k}\}_{k=1}^{\infty}$ form for every i an increasing sequence of measurable sets with

$$\bigcup_{k=1}^{\infty} T_{i,k} = T_i.$$

Hence for every i there is an integer $k(i)$ such that

$$\int_{T_i \setminus T_{i,k(i)}} r(s_0)(t)\,\mu(dt) < \frac{1}{i}.$$

Therefore

$$\int_{T \setminus T_{i,k(i)}} r(s_0)(t) \mu(dt) < \frac{2}{i},$$

$$u_n \rightrightarrows u_0 \quad and \quad r(s_n) \rightrightarrows r(s_0) \quad uniformly \ on \ T_{i,k(i)}$$

and

$$|u_0(t)| < r(s_0)(t) - \frac{1}{k(i)} \quad on \ T_{i,k(i)}.$$

By the uniform convergence there exists a sequence $\{n_i\}_{i=1}^{\infty}$ such that for $n \geq n_i$ we have

$$|u_n(t)| < r(s_n)(t) \quad on \ T_{i,k(i)}.$$

Additionally we may assume that $\{n_i\}_{i=1}^{\infty}$ is strictly increasing. Choose arbitrarily $v_n \in G(s_n)$ and define for $n_i \leq n < n_{i+1}$

$$w_n = u_n \chi_{T_{i,k(i)}} + v_n \chi_{T \setminus T_{i,k(i)}}.$$

Since $G(s_n)$ is decomposable then $w_n \in G(s_n) \subset F$. To the end of the proof we need to show that $w_n \longrightarrow u_0$ in $L^1(T, X)$. But this follows from an observation that for $n_i \leq n < n_{i+1}$ we have the inequalities

$$\|w_n - u_0\|_1 =$$

$$= \int_{T \setminus T_{i,k(i)}} |v_n(t) - u_0(t)| \mu(dt) + \int_{T_{i,k(i)}} |u_n(t) - u_0(t)| \mu(dt) \leq$$

$$\leq \int_{T \setminus T_{i,k(i)}} |r(s_n)(t)| \mu(dt) + \int_{T \setminus T_{i,k(i)}} |u_0(t)| \mu(dt) + \|u_n - u_0\|_1 \leq$$

$$\leq \|r(s_n) - r(s_0)\|_1 + 2 \int_{T \setminus T_{i,k(i)}} |r(s_0)| \mu(dt) + \|u_n - u_0\|_1 \leq$$

$$\leq \|r(s_n) - r(s_0)\|_1 + \|u_n - u_0\|_1 + \frac{4}{i}.$$

This completes the proof. ☐

Now we are going to present a decomposable analogue of the Michael Selection Theorem. The first result of this type belongs to Antosiewicz & Cellina [3]. An abstract setting was found by Fryszkowski [72] and improved by Bressan&Colombo [35]

THEOREM 42 (Bressan-Colombo-Fryszkowski). *Assume that* $P :$ $S \longrightarrow dcl(T, X)$ *is a l.s.c. multifunction. Then* P *admits a continuous selection.*

PROOF. As in two previous selection theorems the proof also goes in three stages. The main difference appears in step I, where we use similar scheme but different arguments.

Step I :   For every $\varepsilon > 0$ we shall construct continuous mappings

$$p : S \longrightarrow L^1(T, X) \quad and \quad \varphi : S \longrightarrow L^1(T, R)$$

such that for each $s \in S$

(9.2) $$\|\varphi(s)\|_1 < \varepsilon$$

and

$$P(s) \cap \{u : |u(t) - p(s)(t)| < \varphi(s)(t)\} \neq \emptyset$$

hold. Fix $\varepsilon > 0$, $s_0 \in S$ and $u_0 \in P(s_0)$. Consider a function

$$\psi_{s_0, x_0} : S \longrightarrow L^1(T, R)$$

given by

$$\psi_{s_0, x_0}(s) = ess \inf \{|u(t) - u_0(t)| : u \in P(s)\}.$$

Obviously

(9.3) $$\psi_{s_0, x_0}(s_0) = 0.$$

Take a multifunction

$$\Phi_{s_0, x_0}(s) = cl \left\{ v \in L^1(T, R) : v(t) > \psi_{s_0, x_0}(s)(t) \quad a.e. \ in \ T \right\}.$$

By the Proposition 57 the mapping $\Phi_{s_0, x_0} : S \longrightarrow clco(L^1(T, R))$ is l.s.c. and, by (9.3), $0 \in \Phi_{s_0, x_0}(s_0)$. Therefore from the Michael Selection Theorem we conclude that Φ_{s_0, x_0} admits a continuous selection $\varphi_{s_0, u_0} : S \longrightarrow L^1(T, R)$ such that

$$\varphi_{s_0, u_0}(s_0) = 0.$$

Consider sets

(9.4) $$V_{s_0, u_0} = \left\{ s \in S : \|\varphi_{s_0, u_0}(s)\|_1 < \frac{\varepsilon}{2} \right\}$$

and observe that V_{s_0, u_0} is an is an open neighbourhood of s_0. Moreover, by the *l.s.c.*, the sets

$$\{V_{s_0, u_0}\}_{s_0 \in S, u_0 \in P(s_0)}$$

form an open covering of the separable metric space S. Therefore there exists a locally finite continuous partition of unity $\{z_n\}_{n=1}^{\infty}$ subordinated to $\{V_{s_0, u_0}\}_{s_0 \in S, u_0 \in P(s_0)}$. Let s_n and u_n be such that

$$z_n^{-1}(0, 1] \subset V_{s_n, u_n}, \quad n = 1, 2, \ldots$$

and set

$$\varphi_n = \varphi_{s_n, u_n} \quad and \quad V_n = V_{s_n, u_n}$$

From (9.4) we conclude that for every $s \in S$ and each $n = 1, 2, ...$

$$z_n(s) \int_T \varphi_n(s)(t)\,\mu(dt) \leq \frac{\varepsilon z_n(s)}{2}.$$

Hence

(9.5) $$\sum_{n=1}^{\infty} z_n(s) \int_T \varphi_n(s)(t)\,\mu(dt) \leq \frac{\varepsilon}{2}.$$

By the Theorem 18 the set T admits a continuous family of partitions $\{A_n(s)\}_{n=1}^{\infty}$ having property that for every $s \in S$ and each $n = 1, 2, ...$

(9.6) $$\left| \sum_{n=1}^{\infty} \int_{A_n(s)} \varphi_n(s)(t)\,\mu(dt) - \sum_{n=1}^{\infty} z_n(s) \int_T \varphi_n(s)(t)\,\mu(dt) \right| < \frac{\varepsilon}{2}.$$

Therefore from (9.5) and 9.6) we have

(9.7) $$\left| \sum_{n=1}^{\infty} \int_{A_n(s)} \varphi_n(s)(t)\,\mu(dt) \right| < \varepsilon.$$

Denote by

$$p(s) = \sum_{n=1}^{\infty} u_n \chi_{A_n(s)}$$

and

$$\varphi(s) = \sum_{n=1}^{\infty} \varphi_n(s)\chi_{A_n(s)}.$$

By (9.7) the condition (9.2) holds. We claim that p and φ are such continuous mappings that for each $s \in S$

$$P(s) \cap \{u : |u(t) - p(s)(t)| < \varphi(s)(t)\} \neq \emptyset.$$

Fix $s \in S$ and let $\Lambda(s) = \{i_1, ..., i_m\}$ be such that $s \in \bigcap_{k=1}^{m} V_{i_k}$. This, in turn, means that

$$p(s) = \sum_{k=1}^{m} u_{i_k}(s)\chi_{A_{i_k}(s)}$$

and

$$\varphi(s) = \sum_{k=1}^{m} \varphi_{i_k}(s)\chi_{A_{i_k}(s)}.$$

Observe that for $k = 1, 2, ..., m$ we have

$$\left\| \varphi_{i_k} (s) \right\|_1 < \frac{\varepsilon}{2}$$

Pick $\overline{u}_k \in P(s)$ in such way that

$$\left| \overline{u}_k (t) - u_{i_k} (s)(t) \right| < \varphi_{i_k} (s)(t) \quad a.e.\ in\ T$$

and take

$$\overline{u} = \sum_{k=1}^{m} \overline{u}_k \chi_{A_{i_k}(s)}.$$

Then, by the decomposability,

$$\overline{u} \in P(s)$$

and

$$\left| \overline{u}(t) - p(s)(t) \right| = \sum_{k=1}^{m} \left| \overline{u}_k (t) - u_{i_k}(t) \right| \chi_{A_{i_k}(s)} <$$

$$< \sum_{k=1}^{m} \varphi_{i_k}(s)(t) \chi_{A_{i_k}(s)}(t) = \varphi(s)(t) \quad a.e.\ in\ T.$$

Step II :   The construction goes by the induction argument.

For $n = 0$ put $P_0(s) = P(s)$.

By the step I with $\varepsilon = \frac{1}{2^1}$ there are continuous $p_1 : S \longrightarrow L^1(T, X)$ and $\varphi_1 : S \longrightarrow L^1(T, R)$ such that for each $s \in S$

$$\left\| \varphi(s) \right\|_1 < \frac{1}{2^1}$$

and

$$P_0(s) \cap \{ u : |u(t) - p_1(s)(t)| < \varphi_1(s)(t) \quad a.e.\ in\ T \} \neq \emptyset.$$

Then by the Proposition 58 the multifunction $P_1 : T \longrightarrow dcl(T, X)$ given by

$$P_1(s) = cl\,\{ P_0(s) \cap \{ u : |u(t) - p_1(s)(t)| < \varphi_1(s)(t) \quad a.e.\ in\ T \} \}$$

is *l.s.c.* .

Assume that we have already constructed continuous

$$p_1, ..., p_n : S \longrightarrow L^1(T, X), \quad \varphi_1, ..., \varphi_n : S \longrightarrow L^1(T, R)$$

and *l.s.c. multifunctions* $P_0, ..., P_{n-1} : S \longrightarrow dcl(T, X)$ such that for every $s \in S$ and for $k = 0, 1, ..., n - 1$ one has

$$\left\| \varphi_n(s) \right\|_1 < \frac{1}{2^n},$$

$$P_{n-1}(s) \cap \{ u : |u(t) - p_n(s)(t)| < \varphi_n(s)(t) \quad a.e.\ in\ T \} \neq \emptyset.$$

and
$$P_{n-1}(s) \subset P_{n-2}(s) \subset \ldots \subset P_0(s).$$
Setting
$$P_n(s) = cl\{P_{n-1}(s) \cap \{u : |u(t) - p_n(s)(t)| < \varphi_n(s)(t) \quad a.e. \ in \ T\}\}$$
we obtain a *l.s.c.* multifunction.

By the *step I* for $\varepsilon = \frac{1}{2^{n+1}}$ there exist: a continuous $p_{n+1} : S \longrightarrow L^1(T, X)$ and $\varphi_{n+1} : S \longrightarrow L^1(T, R)$ such that for each $s \in S$
$$\|\varphi_{n+1}(s)\|_1 < \frac{1}{2^{n+1}}$$
and
$$P_n(s) \cap \{u : |u(t) - p_{n+1}(s)(t)| < \varphi_{n+1}(s)(t) \quad a.e. \ in \ T\} \neq \emptyset.$$
Take
$$P_{n+1}(s) =$$
$$= cl\left(P_n(s) \cap \{u : |u(t) - p_{n+1}(s)(t)| < \varphi_{n+1}(s)(t) \quad a.e. \ in \ T\}\right)$$
and observe that $P_{n+1} : S \longrightarrow dcl(T, X)$ is *l.s.c.* . One can easily check that
$$P_{n+1}(s) \subset P_n(s) \subset \ldots \subset P_0(s)$$
and
$$d(p_{n+1}(s), P_n(s)) \leq \frac{1}{2^{n+1}},$$
what ends the induction step.

Step III: We shall show that $p_n \rightrightarrows p$ and p is the required selection. For any integers $n \geq 0$ and $k \geq 1$ we have

(9.8) $$P_{n+k-1}(s) \subset P_n(s) \subset P(s)$$

and thus
$$d(p_{n+k}(s), P_n(s)) \leq d(p_{n+k}(s), P_{n+k-1}(s)) \leq \frac{1}{2^{n+k}}.$$
Therefore
$$\|p_{n+k}(s) - p_{n+1}(s)\|_1 \leq d(p_{n+k}(s), P_n(s)) + d(p_{n+1}(s), P_n(s)) \leq$$
$$\leq \frac{1}{2^{n+k}} + \frac{1}{2^n} \leq \frac{1}{2^{n-1}},$$
what shows that sequence $\{p_n\}_{n=1}^\infty \subset C(S, L^1(T, X))$ of continuous functions satisfies the uniform Cauchy condition. Thus there is a continuous p such that $p_n \rightrightarrows p$. Such p is the required selection since, by (9.8), we have
$$d(p(s), P(s)) = \lim_{k \longrightarrow \infty} d(p_k(s), P_0(s)) = 0.$$
This completes the proof. $\qquad\qquad\qquad\qquad\qquad\qquad\qquad\qquad\square$

The previous result gives also an answer on a question how many continuous selections posses a *l.s.c.* multifunction $P : S \longrightarrow dcl\,(T, X)$. Especially whether for given $s_0 \in S$ the *l.s.c.* multifunction P admits a continuous selection assuming a prescribed value $u_0 \in P(s_0)$.

COROLLARY 11. *Let* $P : T \longrightarrow dcl\,(T, X)$ *be a l.s.c. multifunction and let* $F \subset S$ *be a closed subset. Consider a continuous function* $p : F \longrightarrow L^1\,(T, X)$ *such that for every* $s \in F$ *we have* $p\,(s) \in P\,(s)$. *Then* p *can be extended to a continuous selection of* P. *In particular for any given* $s_0 \in S$ *and* $u_0 \in P(s_0)$ *there exists a continuous selection* p_{s_0, u_0} *of* P *such that*

(9.9) $p_{s_0, u_0}\,(s_0) = u_0.$

Moreover, for every $s \in S$ *the following representation*

(9.10) $P\,(s) = \{p_{s_0, u_0}\,(s) : s_0 \in S \ \ and \ \ u_0 \in P(s_0)\}$

holds.

PROOF. Consider a multifunction P_F given by

(9.11) $P_F\,(s) = \begin{cases} \{p\,(s)\} & for \ \ s \in F, \\ P\,(s) & for \ \ s \notin F \end{cases}.$

and observe that $P_F : S \longrightarrow dec\,(T, X)$ and it is *l.s.c.*. So the Bressan-Colombo-Fryszkowski Theorem in this case gives a continuous selection which obviously is is a required extension on $P_F \subset P$. Taking $F = \{s_0\}$ we can extend $p_F\,(s_0) = u_0$ to a continuous selection p_{s_0, u_0} of P such that $p_{s_0, u_0}\,(s_0) = u_0$. Since $s_0 \in S$ and $u_0 \in P(s_0)$ are arbitrary one can easily get (9.10).   □

In certain situations instead of (9.10) we can also obtain a Castaing type representation of P through countable many continuous selections. Before we present this type result for multifunctions with decomposable values we need the following:

THEOREM 43. *Let* $P : S \longrightarrow dcl\,(T, X)$ *be a l.s.c. multifunction and assume that there are continuous mappings* $\varphi : S \longrightarrow L^1\,(T, X)$ *and* $r : S \longrightarrow (0, \infty)$ *such that for every* $s \in S$ *the set*

$\Phi\,(s) = P\,(s) \cap B\,(\varphi\,(s), r\,(s)) \neq \emptyset.$

Then the multifunction $\Phi : S \longrightarrow N\,(L^1\,(T, X))$ *admits a continuous selection.*

PROOF. Since the multifunction $\frac{1}{r(s)}\,(P\,(s) - \varphi\,(s))$ is *l.s.c.* and still with closed decomposable values then we may assume that $\varphi \equiv 0$ and $r \equiv 1$. So let

$\Phi\,(s) = P\,(s) \cap B\,(0, 1) \neq \emptyset.$

Fix $s_0 \in S$, $u_0 \in \Phi(s_0)$ and let p_{s_0,u_0} be such continuous selection of P that $p_{s_0,u_0}(s_0) = u_0$. Consider sets

$$(9.12) \qquad V_{s_0,u_0} = \left\{ s \in S : \|p_{s_0,u_0}(s)\|_1 < \frac{1 + \|u_0\|_1}{2} \right\}$$

and observe that V_{s_0,u_0} is an open neighbourhood of s_0. Moreover, by the *l.s.c.*, the family $\{V_{s_0,u_0}\}_{s_0 \in S, u_0 \in P(s_0)}$ is an open covering of the separable metric space S. Therefore there exists a locally finite continuous partition of unity $\{z_n\}_{n=1}^{\infty}$ subordinated to $\{V_{s_0,u_0}\}_{s_0 \in S, u_0 \in P(s_0)}$. Let s_n and u_n be such that

$$z_n^{-1}(0,1] \subset V_{s_n,u_n}, \quad n = 1, 2, \dots$$

and denote by

$$p_n = p_{s_n,u_n} \quad \text{and} \quad V_n = V_{s_n,u_n}$$

From (9.12) we conclude that for every $s \in S$ and each $n = 1, 2, \dots$
(9.13)

$$z_n(s) \|p_n(s)\|_1 = z_n(s) \int_T |p_n(s)(t)| \, \mu(dt) \leq \frac{(1 + \|u_n\|_1) z_n(s)}{2}.$$

Hence

$$\sum_{n=1}^{\infty} z_n(s) \int_T |p_n(s)(t)| \, \mu(dt) \leq \sum_{n=1}^{\infty} \frac{(1 + \|u_n\|_1) z_n(s)}{2}$$

By the Theorem 18 the set T admits a continuous family of partitions $\{A_n(s)\}_{n=1}^{\infty}$ having property that for every $s \in S$ and each $n = 1, 2, \dots$
(9.14)

$$\left| \int_{A_n(s)} |p_n(s)(t)| \, \mu(dt) - z_n(s) \int_T |p_n(s)(t)| \, \mu(dt) \right| < \frac{(1 - \|u_n\|_1) z_n(s)}{2}.$$

Thus

$$\left| \sum_{n=1}^{\infty} \int_{A_n(s)} |p_n(s)(t)| \, \mu(dt) - \sum_{n=1}^{\infty} z_n(s) \int_T |p_n(s)(t)| \, \mu(dt) \right| <$$

$$< \sum_{n=1}^{\infty} \frac{(1 - \|u_n\|_1) z_n(s)}{2}.$$

Adding (9.13) and (9.14) we get

$$(9.15) \qquad \left| \sum_{n=1}^{\infty} \int_{A_n(s)} p_n(s)(t)\,\mu(dt) \right| \leq \sum_{n=1}^{\infty} \int_{A_n(s)} |p_n(s)(t)|\,\mu(dt) < 1.$$

Denote by

$$p(s) = \sum_{n=1}^{\infty} p_n(s)\,\chi_{A_n(s)}$$

and observe that p is continuous selection of P, since such are p_n and A_n. By (9.15) the mapping p is a required selection. $\qquad \square$

REMARK 13. *The reader may have already noticed that the sets $R(s)$ are neither decomposable nor closed. So we have a continouous selection existence result with neither decomposability nor closedness assumptions.*

COROLLARY 12. *Let $P : S \longrightarrow dcl\,(T, X)$ be a l.s.c. multifunction. Fix $s_0 \in S$, $u_0 \in P(s_0)$ and $r > 0$ and let*

$$V = P^- B(u_0, r) = \{s \in S : P(s) \cap B(u_0, r) \neq \emptyset\}.$$

Take a closed $F \subset V$ and consider a multifunction $R : S \longrightarrow N(L^1(T, X))$ given by

$$R(s) = \begin{cases} P(s) \cap B(u_0, r) & for \;\; s \in F, \\ P(s) & for \;\; s \notin F. \end{cases}$$

Then R admits a continuous selection.

PROOF. Observe that $\frac{1}{r(s)}(P(s) - u_0)$ is a l.s.c. multifunction with closed decomposable values and for every $s \in S$ the set

$$\left\{ \frac{1}{r(s)}(P(s) - u_0) \right\} \cap B(0, 1) \neq \emptyset.$$

By the Theorem 43 a multifunction $P_F : F \longrightarrow N(L^1(T, X))$ given by

$$P_F(s) = \left\{ \frac{1}{r(s)}(P(s) - u_0) \right\} \cap B(0, 1)$$

admits a continuous selection $p_F : F \longrightarrow L^1(T, X)$. Take a continuous function $p : F \longrightarrow L^1(T, X)$ given by

$$p(s) = r p_F(s) + u_0$$

and notice that p_F is a continuous selection of $R|_F$ as well as of $P|_F$. By the Corollary 11 it can be extended to a continuous selection p of P and therefore of R. $\qquad \square$

THEOREM 44. *Consider a multivalued mapping* $P : S \longrightarrow dcl\,(T,X)$. *Then P is l.s.c. if and only if there are continuous* $p_n : S \longrightarrow L^1\,(T,X)$, $n = 1, 2, ...,$ *that for every $s \in S$ we have*

(9.16) $P(s) = cl\,\{p_n(s) : n = 1, 2, ...\}.$

PROOF. The idea of the proof is similar to that of the Theorem 4.10. Since the implication $\Longleftarrow$. follows from the Example 6, we have to present $\Longrightarrow$.

Let $\{u_n\}_{n=1}^{\infty}$ be a dense subset in $L^1\,(T,X)$. For every $k = 1, 2, ...$ consider open sets

$$V_{n,k} = P^- B\left(u_n, \frac{1}{k}\right) = \left\{s : P(s) \cap B\left(u_n, \frac{1}{k}\right) \neq \emptyset\right\}.$$

By the choice of $\{u_n\}_{n=1}^{\infty}$ we have

$$S = \bigcup_{n,k=1}^{\infty} V_{n,k}.$$

But in a separable metric space any open set can be decomposed into a countable union of closed sets. Therefore for every $n, k = 1, 2, ...$ there exist closed $F_{n,k,m},\;\; m = 1, 2, ...,$ that

$$V_{n,k} = \bigcup_{m=1}^{\infty} F_{n,k,m}.$$

Consider multifunctions

$$P_{n,k,m}(s) = \begin{cases} cl\,\{P(s) \cap B\left(u_n, \frac{1}{k}\right)\} & for \;\; s \in F_{n,k,m} \\ P(s) & for \;\; s \notin F_{n,k,m} \end{cases}$$

and therefore by the Theorem 43 and the Corollary 12 each $P_{n,k,m} : S \longrightarrow dcl\,(T,X)$ admits a continuous selection $p_{n,k,m} : S \longrightarrow L^1\,(T,X)$. Obviously all $p_{n,k,m}$ are also continuous selections of P. We shall show that for every $s \in S$

(9.17) $P(s) = cl\,\{p_{n,k,m}(s) : n, k, m = 1, 2, ...\}.$

Denote the right-hand side of (9.17) by $R(s)$. Since $R(s) \subset P(s)$ we need to show that $P(s) \subset R(s)$. To see this let us fix $s \in S$, $u \in P(s)$ and $k = 1, 2, ...$ Choose n in such way that

$$u \in P(s) \cap B\left(u_n, \frac{1}{k}\right).$$

Then $s \in V_{n,k}$ and hence there exists m that $s \in F_{n,k,m}(s)$ and $u \in P_{n,k,m}(s)$. Thus for the continuous selection $p_{n,k,m}(s)$ we have

$$\|p_{n,k,m}(s) - u_n\|_1 \leq \frac{1}{k}$$

and therefore

$$\|p_{n,k,m}(s) - u\|_1 < \frac{2}{k}.$$

The latter means nothing else then $d(u, R(s)) < \frac{2}{k}$. But k was chosen arbitrarily, so it implies that $u \in R(s)$. By the choise of u we conclude that $P(s) \subset R(s)$. Renumerating in (9.17) the sequence $p_{n,k,m}$ we obtain (9.16), what ends the proof.   □

2. Carathéodory type selections

Assume now that Ω we have given a complete metric space with a $\sigma - field$ Σ of Lebesgue measurable sets given by a locally finite Radon measure m. On the space $\Omega \times S$ we shall consider the product $\sigma - field$ $\Sigma \otimes \mathcal{B}$. The main object examined in this section is a $l.C.$ multivalued mapping $P : \Omega \times S \longrightarrow dcl\,(T, X)$.

THEOREM 45. *Each l.C. multivalued mapping* $P : \Omega \times S \longrightarrow dcl\,(T, X)$ *admits a Carathéodory type selection.*

PROOF. Take an inreasing family of compact sets $\Omega_m \subset \Omega$, $m = 1, 2, ...$, with $\mathsf{m}\,(\Omega \backslash \Omega_m) < \frac{1}{m}$ such that $P : \Omega_m \times S \longrightarrow dcl\,(T, X)$ is $l.s.c.$. Such sets exist because of the Scorza-Dragoni type Theorem 30. Therefore the Bressan-Colombo-Fryszkowski Theorem yields the existence of a continuous selection of P restricted to each $\Omega_m \times S$. Starting with a continuous selection p on $\Omega_1 \times S \longrightarrow X$ we can consequtively extend it to a continuous selection on each $\Omega_m \times S$ and finally on

$$\Omega \times S = \left(\bigcup_{m=1}^{\infty} \Omega_m \right) \times S.$$

But Ω is a set of full measure. Thus setting $p\,(\omega, s) = 0$ for $(\omega, s) \in (\Omega \backslash \Omega_m) \times S$ we obtain a desired selection.   □

Proceeding exactly in the same way as for the representation (6.3) we obtain the following analogue of the Theorem 33

THEOREM 46. *Let* $P : \Omega \times S \longrightarrow dcl\,(T, X)$ *be a jointly measurable multivalued mapping. Then the following conditions are equivalent:*

    i.   *P is l.C. type;*

    ii. *for every $\varepsilon > 0$ there exists a compact K with $\mathsf{m}\,(\Omega \backslash K) < \varepsilon$ such that* $P : \Omega \times S \longrightarrow dcl\,(T, X)$ *is l.s.c.;*

    iii.   *there exist Carathéodory type functions* $p_n : \Omega \times S \longrightarrow R$, $n \in N$, *such that*

(9.18) $P\,(\omega, s) = cl\,\{p_n\,(\omega, s) : n \in N\}.$

CHAPTER 10

Fixed points property

The fixed point theory can also be developed for single or multivalued self mappings defined on decomposable sets. However a straightforward analogue of the Schauder Theorem can not be provided since each decomposable an compact set reduces to a point. But we can obtain some results assuming compactness of considered mappings.

THEOREM 47. *Let K be a closed decomposable subset of $L^1(T, X)$. Then each compact mapping $f : K \longrightarrow K$ admits a fixed point.*

PROOF. Take $S = clco\{f(K)\}$ and observe that by the Mazur theorem S is compact. Let

$$\psi(s)(t) = ess\inf\{|u(t) - s(t)|\}.$$

For any given $\varepsilon > 0$ and every $s \in S$ consider nonempty sets

$$P_\varepsilon(s) = cl\{u \in K : |u(t) - s(t)| < \psi(s)(t) + \varepsilon \quad a.e. \text{ in } T\}.$$

By the Proposition 58 each mapping

$$P_\varepsilon : S \longrightarrow dcl(T, X)$$

is *l.s.c.* and therefore it admits a continuous selection $p_\varepsilon : S \longrightarrow K$. But for every $s \in f(K)$ we have

$$\psi(s)(t) = 0 \quad a.e. \text{ in } T$$

and hence

(10.1) $$\|p_\varepsilon(s) - s\|_1 \leq \varepsilon.$$

Therefore $f_\varepsilon = f \circ p_\varepsilon : S \longrightarrow f(K) \subset S$ is such continuous function that for every $s \in S$

$$\|f_\varepsilon(s) - s\|_1 \leq \varepsilon.$$

Now the Schauder Theorem yields the existence of $s_\varepsilon \in f(K) \subset S$ such that

(10.2) $$f(p_\varepsilon(s_\varepsilon)) = s_\varepsilon.$$

Passing, if necessary, to a subnet we may assume that $\{s_\varepsilon\}$ is convergent in $f(K) \subset K$, say

$$\lim_{\varepsilon \longrightarrow 0} s_\varepsilon = s_0.$$

Thus by 10.1

$$\lim_{\varepsilon \longrightarrow 0} p_\varepsilon (s_\varepsilon) = s_0.$$

Passing to the limits with $\varepsilon \longrightarrow 0$ in (10.2) one can see that s_0 is a fixed point of f. This completes the proof. $\square$

10.1. Parametrized version. Let $p : \Omega \times K \longrightarrow K$ be a Carathéodory type mapping such that for every $\omega \in \Omega$ the function

$$u \longrightarrow p(\omega, u)$$

is compact. Then the set

(10.3) $\mathcal{F}(\omega) = \{u : u = p(\omega, u)\}$

is nonempty and closed. Moreover $\mathcal{F} : \Omega \longrightarrow cl(L^1(T, X))$ is measurable. Indeed. For every $\varepsilon > 0$ there is a compact set $\Omega_\varepsilon \subset \Omega$ such that $m(\Omega \backslash \Omega_\varepsilon) < \varepsilon$ and $p : \Omega_\varepsilon \times K \longrightarrow K$ is continuous. Therefore it's graph

$$\mathcal{K} = \{(\omega, u, v) \in \Omega_\varepsilon \times K \times K : v = p(\omega, u)\}$$

is closed and then

$$gr\mathcal{F} = proj_{\Omega_\varepsilon \times K} \{(\omega, u, v) \in \mathcal{K} : v = u\}$$

is closed as well. This, in turn, means that $\mathcal{F} : \Omega_\varepsilon \longrightarrow cl(L^1(T, X))$ is *u.s.c.*, what in view of the Scorza-Dragoni Theorem gives measurability of $\mathcal{F}$. Having this we can take a measurable selection of $\mathcal{F}$ obtaining

PROPOSITION 59. *Let $p : \Omega \times K \longrightarrow K$ be such a Carathéodory type function that for every $\omega \in \Omega$ the mapping $u \longrightarrow p(\omega, u)$ is compact. Then there exists a measurable mapping $x : \Omega \longrightarrow K$ such that*

$$x(\omega) = p(\omega, x(\omega))$$

holds a.e. in Ω.

CHAPTER 11

Aumann integrals

1. General properties

We shall assume now that X, X^* are separable Banach spaces, while S and Ω−complete metric spaces. Moreover on Ω we have a $\sigma - field\ \Sigma$ of Lebesgue measurable sets given by a locally finite Radon measure m. Saying about the *(joint)* measurability we mean, respectively, $\mathcal{L} \otimes \Sigma \otimes \mathcal{B}(S)-$, $\mathcal{L} \otimes \Sigma-$ or $\Sigma \otimes \mathcal{B}(S) - measurability$, while a multifunction is a *l.C.* type whenever it is measurable and *l.s.c.* in s.

Let K be a subset of $L^1(T, X)$.

By the Aumann integral of $K \subset L^1(T, X)$ we mean the set

$$\int_T K = \mathcal{A}(K) = \left\{ \int_T u(t)\,\mu(dt) : u \in K \right\}.$$

If $K = K_P$ for some measurable $P : T \longrightarrow N(X)$ then the Aumann integral is also denoted by

$$\int_T K = \mathcal{A}(K) = \int_T P(t)\,\mu(dt) = \int_T P(t)$$

We shall also introduce closed Aumann integral of K. By this term we mean the set

$$cl \int_T K = \mathcal{I}(K) = cl \left\{ \int_T u(t)\,\mu(dt) : u \in K \right\}.$$

The multivalued integrals, called now the Aumann integrals, were introduced in sixties independently by Aumann [11] and Olech [171] for multifunctions $P : T \longrightarrow N(R^l)$ and next extended and examined by many authors for decomposable subsets. They play very important role in differential inclusions and control theory. Shortly speaking their applicability is basically connected with convexity and compactness. In the foundation of that lie the properties of the range of a vector measure. On the other hand the range of a vector measure possessing density $f \in L^1(T, X)$ is the Aumann integral of a measurable multifunction $P_f : T \longrightarrow N(X)$ consisting only of two points $f(t)$ and 0. Therefore many properties of the Aumann integrals can be derived by the use of the Lapunov Theorem. And vice versa, all properties of the Aumann integral have their simpler partners for vector measures.

THEOREM 48. *Let $K \subset L^1(T,X)$ be a closed decomposable set. Then $\mathcal{I}(K) = cl \int_T K$ is convex. If $X = R^l$ then also $\int_T K$ is convex.*

PROOF. Take arbitrary $a, b \in cl \int_T K$ and $\lambda \in I$. We have to show that

$$(11.1) \qquad \lambda a + (1 - \lambda) b \in cl \int_T K.$$

Fix $\varepsilon > 0$ and pick $f, g \in K$ such that

$$(11.2) \qquad \left| a - \int_T f(t) \mu(dt) \right| < \varepsilon \quad and \quad \left| b - \int_T g(t) \mu(dt) \right| < \varepsilon.$$

Consider a measure $m : \Sigma \longrightarrow X^2$ with the density (f, g). Then

$$\lambda \left(\int_T f(t) \mu(dt), \int_T g(t) \mu(dt) \right) \in \mathcal{R}(m).$$

Invoking the Theorem 14 *for the measure m* we conclude that there exists $A \in \Sigma$ such that

$$\left| \lambda \left(\int_T f(t) \mu(dt) \right) - \int_T \chi_A(t) f(t) \mu(dt) \right| < \varepsilon$$

and

$$\left| \lambda \left(\int_T g(t) \mu(dt) \right) - \int_T \chi_A(t) g(t) \mu(dt) \right| < \varepsilon.$$

Notice that the last condition can be equivalently rewritten as

$$\left| (1 - \lambda) \left(\int_T g(t) \mu(dt) \right) - \int_T (1 - \chi_A(t)) g(t) \mu(dt) \right| < \varepsilon.$$

Taking into account the above estimates one can see that the inequalities (11.2) yield

$$\left| \lambda a - \int_T \chi_A(t) f(t) \mu(dt) \right| < 2\varepsilon$$

and

$$\left| (1 - \lambda) b + \int_T (1 - \chi_A(t)) g(t) \mu(dt) \right| < 2\varepsilon.$$

Adding both together we end up with

$$(11.3) \qquad \left| \lambda a + (1 - \lambda) b - \int_T u(t) \mu(dt) \right| < 4\varepsilon,$$

where

$$u = \chi_A f + (1 - \chi_A) g \in K.$$

But $\varepsilon > 0$ was arbitrarily chosen and therefore (11.1) holds.

If $X = R^l$ we can modify (11.2) by taking for arbitrary $a, b \in \int_T K$ such functions $f, g \in K$ that

$$a = \int_T f(t) \mu(dt) \qquad and \qquad b = \int_T g(t) \mu(dt).$$

Now the Lapunov Theorem applied to the measure $m : \Sigma \longrightarrow R^{2l}$ with the density (f, g) leads to the existence of a set $A \in \Sigma$ such that

$$\lambda \left(\int_T (f(t), g(t)) \mu(dt) \right) = \int_T \chi_A(t) (f(t), g(t)) \mu(dt).$$

But this gives

$$\lambda a + (1 - \lambda) b = \int_T u(t) \mu(dt)$$

for

$$u = \chi_A f + (1 - \chi_A) g \in K$$

and completes the proof. $\qquad\qquad\qquad\qquad\qquad\qquad\qquad\qquad\square$

Consider now a decomposable set $K \in dec(T, X)$ given by an $\mathcal{L} -$ *measurable* multifunction $P : T \longrightarrow N(X)$, i.e.

$$K = K_P = \{ u \in L^1(T, X) : u(t) \in P(t) \quad a.e. \ in \ T. \}$$

Together with P we shall also have a multivalued mapping $clcoP$: $T \longrightarrow clco\,(X)$ given by

$$(clcoP)\,(t) = clcoP\,(t)\,.$$

Now we are going to discuss a connection between closed Aumann integrals $\mathcal{I}\,(K) = \mathcal{I}\,(P)$ and $\mathcal{I}\,(clcoP)$. The main tool using for this purpose is the support function. Recall that for $W \subset N\,(X)$ by this term we mean the function $c_W : X^* \longrightarrow \overline{R}$ given by $c_W\,(x^*) = \sup\,\{\langle x^*, u \rangle : u \in W\}$. We also have to notice that $c_W = c_{clcoW}$ and therefore

$$c\,(t, x^*) = c_{P(t)}\,(x^*)$$

is the support function of $clcoP\,(t)\,.$

As a consequnce of the Proposition 39 we shall prove the following:

LEMMA 6. *Let* $K = K_P \in dec\,(T, X)$ *be a decomposable set given by an* $\mathcal{L} - measurable$ *multifunction* $P : T \longrightarrow N\,(X)$. *Then for every* $x^* \in X^*$ *the relation*

$$c_{\mathcal{I}(P)}\,(x^*) = \int_T c\,(t, x^*)\,\mu\,(dt)$$

holds.

PROOF. For any $a = \int_T u\,(t)\,\mu\,(dt) \in \mathcal{A}\,(P)$, where $u \in K_P$, we get

$$\langle x^*, a \rangle = \int_T \langle x^*, u\,(t) \rangle\,\mu\,(dt) \le \int_T c\,(t, x^*)\,\mu\,(dt)\,.$$

Hence

$$c_{\mathcal{I}(P)}\,(x^*) = c_{\mathcal{A}(P)}\,(x^*) \le \int_T c\,(t, x^*)\,\mu\,(dt)\,.$$

For the opposite inequality observe that, by the Corollary 9,

$$clK = K_{clP},$$

where $(clP)\,(t) = cl\,(P\,(t))\,.$ Select a sequence $\{u_n\}_{n \in N} \subset K = K_P$ such that

$$(clP)\,(t) = cl\,\{u_n\,(t)\}\quad a.e.\ in\ T.$$

Then

$$c\,(t, x^*) = c_{P(t)}\,(x^*) = c_{P_0(t)}\,(x^*) = \sup_{n \in N} \langle x^*, u_n\,(t) \rangle\quad a.e.\ in\ T.$$

Modifying $\{u_n\}_{n \in N}$ we can require that

(11.4) $\langle x^*, u_1\,(t) \rangle \le \langle x^*, u_2\,(t) \rangle \le ... \le \langle x^*, u_n\,(t) \rangle \le ...\quad a.e.\ in\ T.$

Indeed. By the decomposability we can replace $u_{n+1} \in K$ with

$$\widetilde{u}_{n+1} = \chi_A u_{n+1} + (1 - \chi_A)\, u_n \in K,$$

where

$$A = \{t \in T : \langle x^*, u_n(t) \rangle \le \langle x^*, u_{n+1}(t) \rangle\}.$$

Thus, by (11.4),

$$c(t, x^*) = \lim \langle x^*, u_n(t) \rangle \quad a.e. \ in \ T.$$

Now using the Fatou Lemma we get

$$\int_T c(t, x^*)\, \mu(dt) \le \liminf \left\langle x^*, \int_T u_n(t)\, \mu(dt) \right\rangle \le c_{\mathcal{I}(P)}(x^*),$$

as desired. This ends the proof. $\qquad\square$

From the Lemma 6 we conclude

PROPOSITION 60. *Let* $K = K_P \in dec\,(T, X)$ *be a decomposable set given by an* $\mathcal{L} - measurable$ *multifunction* $P : T \longrightarrow N(X)$. *Then* $\mathcal{I}(K) = \mathcal{I}(P) = \mathcal{I}(clcoP)$.

PROOF. Obviously $\mathcal{I}(P) \subset \mathcal{I}(clcoP)$ and both sets are closed and convex. Therefore to see that they coincide we need only to show that their support functions are the same. For this purpose take any $a \in \mathcal{I}(clcoP) = cl \int_T clcoP(t)\, \mu(dt)$. Then $a = \lim_{n \longrightarrow \infty} a_n$ with

$$a_n = \int_T u_n(t)\, \mu(dt)$$

for some integrable u_n such that $u_n(t) \in clcoP(t)$ a.e. in T. Therefore for every $x^* \in X^*$ and $n \in N$ we have

$$\langle x^*, u_n(t) \rangle \le c_{clcoP(t)}(x^*) = c_{P(t)}(x^*) \quad a.e. \ in \ T.$$

Thus $\langle x^*, a \rangle = \lim_{n \longrightarrow \infty} \left\langle x^*, \int_T u_n(t)\, \mu(dt) \right\rangle \le \int_T c_{P(t)}(x^*)\, \mu(dt) = c_{\mathcal{I}(P)}(x^*)$
and so

$$c_{clco\mathcal{I}(P)}(x^*) \le c_{\mathcal{I}(P)}(x^*).$$

Since the opposite inequality is obvious the proof is completed. $\qquad\square$

1.1. Aumann integrals in R^l. As we have already seen the Aumann integrals $\int_T K$ of decomposable sets K in finitely dimensional spaces are convex. This property is precisely a consequence of the Lapunov theorem for nonatomic vector measures. But the range of a measure is also compact and up-to now we did not examine an influence of this information on the Aumann integrals. This question is strictly tightened with the Olech's Theorem 26 concerning lexicographical order and extremal selections. We begin with the following result

PROPOSITION 61 (Olech Lemma [**171**]). *Let $K \in dec\,(T, R^l)$ and take an extreme point $e \in \mathcal{I}(K) = cl \int_T K$. Then for each $\varepsilon > 0$ there exists a $\delta = \delta\,(\varepsilon) > 0$ such that the condition*

$$(11.5) \qquad \int_T u\,(t)\,\mu\,(dt), \int_T v\,(t)\,\mu\,(dt) \in B\,(e, \delta),$$

whenever $u, v \in K$, implies

$$\|u - v\|_1 < \varepsilon.$$

PROOF. Without loss of generality we may assume that $e = 0$.

Take $r = \frac{\varepsilon}{4l}$ and observe that, by the extremality, 0 is not a member of $clco\,\{\mathcal{I}\,(K) \setminus B\,(0, r)\}$. Thus $clco\,\{\mathcal{I}\,(K) \setminus B\,(0, r)\}$ can be strongly separated from 0. The latter means that there is a unit vector $p \in R^l$ such that for some $\delta > 0$ the inequality

$$(11.6) \qquad \alpha = \sup\,\{\langle p, a \rangle : a \in \mathcal{I}\,(K) \setminus B\,(0, r)\} < -\delta$$

holds.

We shall show that such choice of δ implies the desire condition. First of all let us notice that

$$(11.7) \qquad \qquad \mathcal{I}\,(K) \cap B\,(0, \delta) \subset B\,(0, r).$$

Indeed. If for some $u \in K$

$$\int_T u\,(t)\,\mu\,(dt) \in \{\mathcal{I}\,(K) \cap B\,(0, \delta)\} \setminus B\,(0, r)$$

then by (11.6)

$$\left| \int_T u\,(t)\,\mu\,(dt) \right| \geq - \int_T \langle p, u\,(t) \rangle\,\mu\,(dt) \geq -\alpha > \delta,$$

what is impossible.

Fix any $a, b \in \mathcal{I}(K) \cap B(0, \delta)$ and let

$$a = \int_T u(t)\, \mu(dt) \quad \text{and} \quad b = \int_T v(t)\, \mu(dt)$$

for some $u, v \in K$. For arbitrary measurable $A \in \mathcal{L}$ put

$$c = \int_A (v(t) - u(t))\, \mu(dt).$$

Observe that

$$a + c = \int_T u(t)\, \mu(dt) + \int_A (v(t) - u(t))\, \mu(dt) =$$

$$= \int_T (v(t)\chi_A + (1 - \chi_A)\, u(t))\, \mu(dt) \in \mathcal{I}(K)$$

and similarly

$$b - c = \int_T v(t)\, \mu(dt) - \int_A (v(t) - u(t))\, \mu(dt) =$$

$$= \int_T (u(t)\chi_A + (1 - \chi_A)\, v(t))\, \mu(dt) \in \mathcal{I}(K).$$

If both

$$a + c, b - c \in \mathcal{I}(K) \setminus B(0, r)$$

then

$$\frac{1}{2}(a + b) = \frac{1}{2}[(a + c) + (b - c)] \in clco\{\mathcal{I}(K) \setminus B(0, r)\},$$

what in view of (11.6) yields

$$|a + b| \geq -2\left\langle p, \frac{1}{2}(a + b)\right\rangle > 2\delta$$

and contradicts with the choice of a and b. Thus at least one of vectors $a + c$ and $b - c$ comes from $B(0, r)$. On the other hand, by (11.7), both $a, b \in B(0, r)$ and therefore $c \in B(0, 2r)$. So we have proved that for each measurable $A \in \mathcal{L}$

$$\left| \int_A (u(t) - v(t))\, \mu(dt) \right| < 2r$$

and this in turn gives for each $i - th$ coordinate, $i \in \{1, ..., l\}$, an estimate

$$\left| \int_A (u_i(t) - v_i(t)) \mu(dt) \right| < 2r.$$

Fix $i = 1, ..., l$. Introducing $A_i = \{t : u_i(t) - v_i(t) \geq 0\}$ we have

$$\int_T |u_i(t) - v_i(t)| \mu(dt) =$$

$$= \left| \int_{A_i} (u_i(t) - v_i(t)) \mu(dt) \right| + \left| \int_{T \backslash A_i} (u_i(t) - v_i(t)) \mu(dt) \right| < 4r.$$

So finally

$$\|u - v\|_1 \leq \sum_{i=1}^{l} \int_T |u_i(t) - v_i(t)| \mu(dt) < 4rl = \varepsilon,$$

what we were suppose to prove.   □

The Olech Lemma has several important implications for the Aumann integral of closed decomposable sets. Some of them are explained in the following

THEOREM 49. *Let $K = K_P$ be a closed decomposable set given by a measurable $P : T \longrightarrow cl\left(R^l\right)$. Then*

   i. *each extremal point $e \in \mathcal{I}(K)$ can be uniquely represented as*

$$e = \int_T k(t) \mu(dt),$$

where $k(t) \in extP(t)$ a.e. in T.

   ii. *if $\mathcal{I}(K)$ is compact then $\mathcal{I}(K) = \int_T K$. More precisely, for each $x \in \mathcal{I}(K)$ there exist: $k_i(t) \in extP(t)$ a.e. in T, $i = 0, 1, ..., l$, and a partition $\{A_i\}_{i=0}^{l}$ such that*

$$x = \int_T \left(\sum_{i=0}^{l} \chi_{A_i}(t) k_i(t) \right) \mu(dt).$$

PROOF. i.   Take an extremal point $e \in ext\mathcal{I}(K)$ and let $\{u_n\}_{n \in N} \subset K$ be a sequence with the property that

$$\int_T u_n(t) \mu(dt) \longrightarrow e.$$

The Olech Lemma says that $\{u_n\}_{n\in N}$ is a Cauchy sequence in $L^1\left(T, R^n\right)$ and therefore convergent to $k \in K$. Hence

$$e = \int_T k\left(t\right) \mu\left(dt\right) \in \int_T K.$$

This representation is unique. Indeed. If for some $u, v \in K$

$$\int_T u\left(t\right) \mu\left(dt\right) = \int_T v\left(t\right) \mu\left(dt\right) = e$$

then for each $\varepsilon > 0$ we have $\|u - v\|_1 < \varepsilon$ since both integrals are in $B\left(e, \delta\left(\varepsilon\right)\right)$. It remains to verify that

$$k\left(t\right) \in extP\left(t\right) \quad a.e. \ in \ T.$$

Actually we shall prove something more. Namely, if for certain basis $\mathcal{E} = \{\mathbf{a}_1, \mathbf{a}_2, ..., \mathbf{a}_l\} \in \Xi$, we have

$$e = e\left(\int_T K, \mathcal{E}\right),$$

then

(11.8) $$k\left(t\right) = e\left(P\left(t\right), \mathcal{E}\right) \quad a.e. \ in \ T.$$

The basis $\mathcal{E}$ induces a convex cone $\times$ having property that

$$\times \cup \left(-\times\right) = R^l \quad and \quad \times \cap \left(-\times\right) = \{0\}.$$

Fix any $u \in K$ and notice that

$$\int_T \left(k\left(t\right) - u\left(t\right)\right) \mu\left(dt\right) = e - \int_T u\left(t\right) \mu\left(dt\right) \in \times$$

Denote by

$$A = \{t : k\left(t\right) - u\left(t\right) \in \left(-\times\right) \backslash \{0\}\}$$

and take

$$w = u\chi_A + \left(1 - \chi_A\right) k \in K.$$

Then

$$k - w = \left(k - u\right)\chi_A \in -\times,$$

while

$$\int_T \left(k\left(t\right) - w\left(t\right)\right) \mu\left(dt\right) \in \times.$$

Such a situation can occur only when the set A is of measure 0 and this, in turn, means that

$$k\left(t\right) - u\left(t\right) \in \times \quad a.e. \ in \ T.$$

But $u \in K$ was arbitrarily chosen, thus

$$k\left(t\right) = e\left(P\left(t\right), \mathcal{E}\right) \quad a.e. \ in \ T,$$

what shows (11.8).

    ii. By Theorem (5) any given point $x \in \mathcal{I}\left(K\right)$ can be represented as

$$x = \sum_{i=0}^{l} \lambda_i e_i,$$

with $e_i \in ext\mathcal{I}\left(K\right)$ and $\lambda_i \geq 0$ such that $\sum_{i=0}^{l} \lambda_i = 1$. Write each e_i as

$$e_i = \int_{T} k_i\left(t\right) \mu\left(dt\right),$$

where $k_i\left(t\right) \in extP\left(t\right)$ a.e. in T. Consider a measure $m : \mathcal{L} \longrightarrow R^{l(l+1)}$ with the density $(k_0, ..., k_l)$ and let $\{A_\alpha\}_{\alpha \in I}$ be a segment for m. Denote $z_i = \sum_{j=0}^{i} \lambda_j$ and take

$$A_i = A_{z_i} \backslash A_{z_{i-1}}.$$

Observe that for each i by construction we have

$$\int_{A_i} k_i\left(t\right) \mu\left(dt\right) = \lambda_i \int_{T} k_i\left(t\right) \mu\left(dt\right).$$

Hence

$$x = \sum_{i=0}^{l} \lambda_i \int_{T} k_i\left(t\right) \mu\left(dt\right) = \sum_{i=0}^{l} \int_{A_i} k_i\left(t\right) \mu\left(dt\right)$$

$$= \int_{T} \left(\sum_{i=0}^{l} \chi_{A_i}\left(t\right) k_i\left(t\right) \right) \mu\left(dt\right),$$

what gives the desired formula. $\qquad\qquad\qquad\qquad\qquad\qquad\qquad\square$

REMARK 14. *If $K = K_P$ is a closed decomposable and integrably bounded set given by a measurable $P : T \longrightarrow cl\left(R^l\right)$ then*

$$\mathcal{I}\left(K\right) = \int_{T} K = \int_{T} extP\left(t\right).$$

2. Aumann integrals of jointly measurable multifuctions

The main object examined in this section is the Aumann integral of a multivalued mapping $P : T \times \Omega \times S \longrightarrow N(X)$, $P : T \times \Omega \longrightarrow N(X)$ or $P : \Omega \times S \longrightarrow dec(T, X)$. We shall also emphasize that in the presented below theory it is the same to consider mappings $P : T \times \Omega \times S \longrightarrow N(X)$ or $K = K_P : \Omega \times S \longrightarrow dec(T, X)$.

Denote by

$$\mathcal{A}(\omega, s) = \int_T P(t, \omega, s) \quad and \quad \mathcal{I}(\omega, s) = cl \int_T P(t, \omega, s).$$

Observe that by the Theorem 14 the multifunction $\mathcal{I} : \Omega \times S \longrightarrow clco(X)$. We are going to discuss problems of preserving measurability or $l.s.c.$ properties of P through the Aumann integrals and vice versa - when and which selections of the Aumann integrals can be possibly regained by suitable selections of the integrand multifunction. For this purpose it is enough to examine the mapping $\mathcal{A}$ since the regularity of $\mathcal{I}$ is of the same type as possess $\mathcal{A}$. We begin with the following

PROPOSITION 62. *Assume that* $P : T \times \Omega \times S \longrightarrow N(X)$ *is a jointly measurable multifuction and integrably bounded. Then* $\mathcal{I} : \Omega \times S \longrightarrow clco(X)$ *is again jointly measurable one. Moreover, if* P *is a l.C. type multivalued mapping then* $\mathcal{I}$ *is again a l.C. type one.*

PROOF. We first show that the Aumann integral preserves $\Sigma \otimes \mathcal{B}(S) - measurability$. In view of the Proposition 40 we have to show that the function $c(s, x^*) = c_{\mathcal{I}(s)}(x^*)$ is $\Sigma - measurable$ in s for every $x^* \in X^*$. From the Proposition 6 we know that

$$c_{\mathcal{I}(\omega, s)}(x^*) = \int_T c(t, \omega, s, x^*) \mu(dt),$$

where $c(t, \omega, s, x^*) = c_{P(t, \omega, s)}(x^*)$. But for every x^* the mapping $(t, \omega, s) \longrightarrow c(t, \omega, s, x^*)$ is $\mathcal{L} \otimes \Sigma \otimes \mathcal{B}(S) - measurable$. Therefore integrating it with respect to t we obtain, by the Proposition 43, required measurability.

Passing to $l.C.$ we actually to show that if $P : S \longrightarrow dec(T, X)$ is $l.s.c.$ mapping then $s \longrightarrow \mathcal{A}(s) = \int_T P(s)$ is $l.s.c.$ as well. To do that pick $s_0, x_0 \in \mathcal{A}(s_0)$ and $s_\alpha \longrightarrow s_0$. We have to find points $x_\alpha \in \mathcal{A}(s_\alpha)$ such that $x_\alpha \longrightarrow x_0$. But $x_0 = \int_T u_0(t) \mu(dt)$ for some $u_0 \in P(s_0)$. By the $l.s.c.$ of P there are $u_\alpha \in P(s_\alpha)$ such that $u_\alpha \longrightarrow u_0$. Then

$$x_\alpha = \int_T u_\alpha(t) \mu(dt) \longrightarrow \int_T u_0(t) \mu(dt) = x_0. \qquad \square$$

CHAPTER 12

Selections of Aumann integrals

If $P : \Omega \times S \longrightarrow dec\,(T, X)$ is a jointly measurable or *l.C. mapping* then the Proposition 62 preserves that property for the Aumann integral $\mathcal{I} : \Omega \times S \longrightarrow clco\,(X)$ given by

$$\mathcal{I}\,(\omega, s) = \mathcal{I}\,(P\,(\omega, s)) = cl \int_T P\,(\omega, s)\,.$$

Moreover, any jointly measurable or Carathéodory type selection $p\,(\omega, s) \in P\,(\omega, s)$ produces, respectively, a measurable or Carathéodory type selection $a\,(\omega, s) = \int_T p\,(\omega, s)\,(t)\,\mu\,(dt)$ of $\mathcal{I}\,(\omega, s)$. So we have, in this case, "a short proof" of the Kuratowski & Ryll-Nardzewski Theorem as well as Michael Selection Theorem and the Theorem 32. The described above situation can be in a sense reversed. Namely, any selection of Aumann integrals can be precisely or "uniformly" regained by a selection of the integrand multifunction. These effects are however different for *l.s.c.* mappings and for measurable ones. The possibility of recovering integrand functions occures to be a consequence of similar properties for families of vector measures and the Lapunov Theorem. On the other hand each given function $p : \Omega \times S \longrightarrow L^1\,(T, X)$ produces a multifunction $P : \Omega \times S \longrightarrow dcl\,(T, X)$ by setting

$$P\,(\omega, s) = dec\,\{p\,(\omega, s)\,, 0\}\,.$$

For such objects we have

$$\mathcal{I}\,(P\,(\omega, s)) = cl \left\{ \int_A p\,(\omega, s)\,(t)\,\mu\,(dt) : A \in \mathcal{L} \right\} = \mathcal{R}\,(\mu\,(\omega, s))\,,$$

where $\mu\,(\omega, s)$ is the vector measure with the density $p\,(\omega, s)\,.$

1. Continuous selections

Regaining selections of Aumann integrals by selections of integrand multifunctions we begin with the continuous case.

THEOREM 50. *Let $P : S \longrightarrow dec\,(T, X)$ be l.s.c. and take a continuous selection $a\,(s) \in \mathcal{I}\,(s) = cl \int_T P\,(s)$. Then for every continuous $\varepsilon : S \longrightarrow R^+$ there exists a continuous selection $p\,(s) \in P\,(s)$ such that for every $s \in S$ we have*

$$\left| a\,(s) - \int_T p\,(s)\,(t)\,\mu\,(dt) \right| < \varepsilon\,(s).$$

PROOF. Replacing for given continuous $\varepsilon : S \longrightarrow R^+$ the multifunction P with $\frac{1}{\varepsilon} P$ and a with $\frac{1}{\varepsilon} a$ we may require that

$$\varepsilon\,(s) \equiv 1.$$

Fix $s_0 \in S$ and choose $u_0 \in P\,(s_0)$ such that

$$\left| a\,(s_0) - \int_T u_0\,(t)\,\mu\,(dt) \right| < \frac{1}{2}.$$

By the Collorary 21 there exists such continuous selection p_{s_0,u_0} of P that

$$p_{s_0,u_0}\,(s_0) = u_0$$

Consider the set

$$V_{s_0,u_0} = \left\{ s \in S : \left| a\,(s) - \int_T p_{s_0,u_0}\,(s)\,(t)\,\mu\,(dt) \right| < \frac{1}{2} \right\}.$$

and observe that $\{V_{s_0,u_0}\}_{s_0 \in S, u_0 \in P(s_0)}$ form an open covering of the separable space S. So one can select a countable and locally finite subcovering and a countable, locally finite partition of unity $\{\varphi_n\,(\cdot)\}_{n=1}^{\infty}$ subordinated to it. Let $V_n = V_{s_n,u_n}$ be such sets that for each $n = 1, 2, \ldots$ $supp\varphi_n \subset V_n$. Therefore for $n = 1, 2, \ldots$ the following inequalities are satisfied:

$$\left| \varphi_n\,(s)\,a\,(s) - \varphi_n\,(s) \int_T p_n\,(s)\,(t)\,\mu\,(dt) \right| \le \frac{1}{2}\varphi_n\,(s),$$

where $p_n = p_{s_n,u_n}$. Thus

$$(12.1) \qquad \left| a\,(s) - \sum_{n=1}^{\infty} \varphi_n\,(s) \int_T p_n\,(s)\,(t)\,\mu\,(dt) \right| \le \frac{1}{2}.$$

Invoking the Theorem 18 we can choose a continuous family of partitions $\{A_n(s)\}_{n=1}^{\infty}$ of the set T having property that for every $s \in S$

$$(12.2) \quad \left| \sum_{n=1}^{\infty} \int_{A_n(s)} p_n(s)(t)\,\mu(dt) - \sum_{n=1}^{\infty} \varphi_n(s) \int_T p_n(s)(t)\,\mu(dt) \right| < \frac{1}{2}.$$

So from (12.1) and (12.2) we have

$$(12.3) \quad \left| a(s) - \sum_{n=1}^{\infty} \int_{A_n(s)} p_n(s)(t)\,\mu(dt) \right| < 1.$$

Denoting by

$$p(s) = \sum_{n=1}^{\infty} p_n(s)\,\chi_{A_n(s)}$$

we get a required continuous selection p of P, since such are $p_n = p_{s_n, u_n}$. This ends the proof. □

2. Continuous selections of the range of vector measures

The consideration from the previous section we shall apply to obtain continuous selections for the range of a continuous family of vector measures. For given continuous mapping $f : S \longrightarrow L^1(T, X)$ we shall consider vector measures $m(s)$ with the density $f(s)$. In such situation the range of $m(s)$ we shall also call the range of $f(s)$.

With any $f(s)$ we can assign a decomposable set

$$P_f(s) = dec\{f(s), 0\}.$$

Then $P_f : S \longrightarrow dcl(T, X)$ is a l.s.c. mapping and one check that

$$\mathcal{R}_f(s) = \mathcal{R}(m(s)) = cl \left\{ \int_A f(s)(t)\,\mu(dt) : A \in \mathcal{L} \right\} = \mathcal{I}(P_f(s)).$$

THEOREM 51. *Let* $f : S \longrightarrow L^1(T, X)$ *be a continuous mapping and denote by* $\mathcal{R}(s) = \mathcal{R}_{\tilde{f}}(s)$ *the range of* $\tilde{f}(s) = (f(s), 1) : S \longrightarrow L^1(T, X \times R)$. *Let* $(r(s), r_0(s))$ *be a continuous selection of* $\mathcal{R}(s)$. *Then for every* $\varepsilon > 0$ *there exists a continuous function* $A : S \longrightarrow \mathcal{L}$ *such that for every* $s \in S$ *we have*

$$\left| r(s) - \int_{A(s)} f(s)(t)\,\mu(dt) \right| < \varepsilon$$

and

$$|r_0(s) - \mu[A(s)]| < \varepsilon.$$

PROOF. The Theorem 50 yields, for every $\varepsilon > 0$, the existence of continuous selection

$$p(s) \in P_{\widetilde{f}}(s) = dec\left\{\widetilde{f}(s), 0\right\}$$

such that

$$\left|(r(s), r_0(s)) - \int\limits_{T} p(s)(t)\,\mu(dt)\right| < \varepsilon.$$

But $p(s) = \left(f(s)\chi_{A(s)}, \chi_{A(s)}\right)$ for some $A(s)$. Thus $A : S \longrightarrow \mathcal{L}$ is such continuous mapping that for every $s \in S$

$$\left|(r(s), r_0(s)) - \int\limits_{A(s)} (f(s)(t), 1)\,\mu(dt)\right| < \varepsilon,$$

as required.   □

3. Measurable selections

We shall now consider a mapping $P : T \times \Omega \longrightarrow c(X)$ which is $\mathcal{L} \otimes \Sigma - measurable$. About P we shall assume that for every $\omega \in \Omega$ the mapping $P(.,\omega)$ is integrably bounded by $m_\omega \in L^1(T)$. Taking the Aumann integral of $P(t,\omega)$ with respect to t we have a compact and convex set

$$\mathcal{A}(\omega) = \int\limits_{T} P(t,\omega)$$

and $\mathcal{A} : \Omega \longrightarrow cc(X)$ is a $\Sigma - measurable$ multifunction. Moreover any $\mathcal{L} \otimes \Sigma - measurable$ selection $p(t,\omega) \in P(t,\omega)$ determines a $\Sigma - measurable$ selection $a(\omega) = \int_T p(t,\omega)\,\mu(dt)$ of $\mathcal{A}$. The described above situation can be in a sense reversed. We discuss first this phenomenon for the range of a measurable family of vector measures. Let $m : \Omega \longrightarrow \mathcal{M}_a(T, R^l)$ be a measurable mapping provided each $m(\omega)$ is a vector measure with the density $f(\omega)$ and $f : \Omega \longrightarrow L^1(T, R^l)$ is measurable. To any density function $f(\omega)$ we can assign a decomposable set

$$P_f(\omega) = dec\{f(\omega), 0\}.$$

Then $P_f : \Omega \longrightarrow dcl\left(T, X \times R^l\right)$ is a measurable mapping and one can check that

$$R\left(\omega\right) = R\left(m\left(\omega\right)\right) = \left\{\int_A f\left(\omega\right)\left(t\right)\mu\left(dt\right) : A \in \mathcal{L}\right\} = \mathcal{I}\left(P_f\left(\omega\right)\right).$$

LEMMA 7. *Consider measurable family of nonatomic vector measures $\widetilde{m} = (m, m_0) : \Omega \longrightarrow \mathcal{M}\left(T, X \times R^l\right)$ with $m : \Omega \longrightarrow \mathcal{M}\left(T, X\right)$ and $m_0 : \Omega \longrightarrow \mathcal{M}\left(T, R^l\right)$. Then for every measurable $\varepsilon : \Omega \longrightarrow R^+$ there exists a measurable mapping $A : \Omega \longrightarrow \mathcal{L}$ such that*

$$\left|m\left(\omega\right)\left(A\left(\omega\right)\right) - \frac{1}{2}m\left(\omega\right)\left(T\right)\right| \leq \varepsilon\left(\omega\right)$$

and

$$m_0\left(\omega\right)\left(A\left(\omega\right)\right) = \frac{1}{2}m_0\left(\omega\right)\left(T\right).$$

PROOF. Endow $\mathcal{L}$ with the metric

$$d_1\left(A, B\right) = \|\chi_A - \chi_B\|_1 = \mu\left(A \triangle B\right)$$

and recall that $(\mathcal{L}, d_1)$ is a separable and complete metric space. For each $\omega \in \Omega$ denote by

$$R\left(\omega\right) = \left\{A \in \mathcal{L} : \begin{array}{l} \left|m\left(\omega\right)\left(A\right) - \frac{1}{2}m\left(\omega\right)\left(T\right)\right| \leq \varepsilon\left(\omega\right) \\ and \quad m_0\left(\omega\right)\left(A\right) = \frac{1}{2}m_0\left(\omega\right)\left(T\right) \end{array}\right\}.$$

Obviously each $R\left(\omega\right)$ is nonempty and closed. We claim that $R : \Omega \longrightarrow cl\left(\mathcal{L}\right)$ is a measurable multifunction. To see this choose for each $\tau > 0$ a compact set $\Omega_\tau \subset \Omega$ such that $\mathsf{m}\left(\Omega \backslash \Omega_\tau\right) < \tau$ and both $\widetilde{m} : \Omega_\tau \longrightarrow \mathcal{L}$ and $\varepsilon : \Omega_\tau \longrightarrow R^+$ are continuous. Therefore $R_{|\Omega_\tau}$ has closed graph what in turn gives the measurability of R. Now the Kuratowski & Ryll-Nardzewski makes a deal. $\square$

Equipped with the previous Lemma we can exactly repeat a construction provided in the Theorem 15 obtaining:

PROPOSITION 63. *Let $\widetilde{m} = (m, m_0) : \Omega \longrightarrow \mathcal{M}\left(T, X \times R^l\right)$ be a measurable family of nonatomic vector measures with $m\left(\omega\right) \in \mathcal{M}_a\left(T, X\right)$ and $m_0\left(\omega\right) : \Sigma \longrightarrow R^l$. Then for every $\varepsilon > 0$ there exists a Carathéodory type mapping $A : \Omega \times I \longrightarrow \mathcal{L}$ such that for each $\omega \in \Omega$ the family $\left\{A\left(\omega, \alpha\right)\right\}_{\alpha \in I}$ is an $\varepsilon-$segment for $m\left(\omega\right)$ and a segment for $m_0\left(\omega\right)$.*

Now we are prepared to prove some existence results concerning measurable selections of Aumann integrals. The simplest situation is for $R^l - valued$ multifunctions. Namely the following holds:

THEOREM 52 (Artstein [6]). *Let* $P : T \times \Omega \longrightarrow cl\left(R^l\right)$ *be an* $\mathcal{L} \otimes \Sigma-$
measurable multifunction and assume that $P\left(\cdot, \omega\right)$ *is for each* $\omega \in \Omega$
integrably bounded by $\beta_\omega \in L^1\left(T\right)$. *Then for each* $\Sigma - measurable$
selection

$$a\left(\omega\right) \in \mathcal{A}\left(\omega\right) = \int_T P\left(t, \omega\right).$$

there exists an $\mathcal{L} \otimes \Sigma - measurable$ *selection* $p\left(t, \omega\right) \in P\left(t, \omega\right)$ *such
that for every* $\omega \in \Omega$ *we have*

$$a\left(\omega\right) = \int_T p\left(t, \omega\right) \mu\left(dt\right).$$

PROOF. Since $\mathcal{A} : \Omega \longrightarrow cc\left(R^l\right)$ is measurable then using the
Theorem 27 we can represent any given measurable selection $a\left(\omega\right) \in \mathcal{A}\left(\omega\right)$ in the form

$$a\left(\omega\right) = \sum_{i=0}^{l} \lambda_i\left(\omega\right) e_i\left(\omega\right),$$

with measurable $\lambda_i : \Omega \longrightarrow [0,1]$ and $e_i\left(\omega\right) \in ext\mathcal{A}\left(\omega\right)$ such that
$\sum_{i=0}^{l} \lambda_i\left(\omega\right) = 1$. But each $e_i\left(\omega\right) = \int_T p_i\left(t, \omega\right) \mu\left(dt\right)$ with measurable
selections $p_i\left(t, \omega\right)$ of $P\left(t, \omega\right)$. By the Proposition 63 there exists a
Carathéodory type mapping $A : \Omega \times I \longrightarrow \mathcal{L}$ such that for every $\omega \in \Omega$
the family $\left\{A\left(\omega, \alpha\right)\right\}_{\alpha \in I}$ is a segment for each $p_i\left(t, \omega\right)$, $i = 0, 1, ..., l$.
The latter means that for each $\alpha \in I$ and $i = 0, 1, ..., l$ the relations

$$(12.4) \qquad \int_{A(\omega,\alpha)} p_i\left(t, \omega\right) \mu\left(dt\right) = \alpha \int_T p_i\left(t, \omega\right) \mu\left(dt\right) = \alpha e_i\left(\omega\right)$$

hold. Denote

$$z_i\left(\omega\right) = \sum_{j=0}^{i} \lambda_j\left(\omega\right)$$

and let

$$B_i\left(\omega\right) = A\left(\omega, z_i\left(\omega\right)\right) \backslash A\left(\omega, z_{i-1}\left(\omega\right)\right).$$

By construction $\left\{B_i\left(\omega\right)\right\}_{i=1}^{n}$ is a measurable family of partitions of T
such that

$$(12.5) \qquad \int_{B_i(\omega)} p_i\left(t, \omega\right) \mu\left(dt\right) = \lambda_i\left(\omega\right) \int_T p_i\left(t, \omega\right) \mu\left(dt\right) = \lambda_i\left(\omega\right) e_i\left(\omega\right)$$

and hence

$$a(\omega) = \sum_{i=0}^{l} \lambda_i(\omega) e_i(\omega) = \int_T p(t,\omega)\mu(dt)$$

with measurable

$$p(t,\omega) = \sum_{i=0}^{l} \chi_{B_i(\omega)}(\omega) p_i(t,\omega).$$

This ends the proof. □

For $\mathcal{L} \otimes \Sigma$–measurable multifunction $P : T \times \Omega \longrightarrow cl(X)$ assuming the values in the Banach space X the situation is similar, however we can regain measurable selections only uniformly. We shall explain this in terms of the mapping

$$K(\omega) = K_{P(\omega,\cdot)}.$$

THEOREM 53. *Let $K : \Omega \longrightarrow dcl(T,X)$ be a Σ – measurable multifunction and assume that $K(\omega)$ is for each $\omega \in \Omega$ integrably bounded by $\beta_\omega \in L^1(T)$. Fix a Σ – measurable selection*

$$a(\omega) \in cl \int_T K(\omega).$$

Then for every measurable $\varepsilon : \Omega \longrightarrow R^+$ one can find a Σ–measurable selection $k(\omega) \in K(\omega)$ such that for every $\omega \in \Omega$ the inequality

$$\left| a(\omega) - \int_T k(\omega)\mu(dt) \right| \leq \varepsilon(\omega)$$

holds.

PROOF. Fix measurable $\varepsilon : \Omega \longrightarrow R^+$ and for every $\omega \in \Omega$ consider the sets

$$P(\omega) = \left\{ u \in K(\omega) : \left| a(\omega) - \int_T u(t)\mu(dt) \right| \leq \varepsilon(\omega) \right\}.$$

Obviously each $P(\omega)$ is nonempty and closed subset of $L^1(T,X)$. We shall show that $P : \Omega \longrightarrow cl(L^1(T,X))$ is measurable. By the Castaing & Jacobs Theorem 41 we may find in Ω an increasing family of compact subsets $\{\Omega_n\}_{n\in N}$ with $m(\Omega \backslash \Omega_n) \xrightarrow[n\to\infty]{} 0$ such that each $K : \Omega_n \longrightarrow dcl(T,X)$ has the closed graph, while $a : \Omega_n \longrightarrow X$ and $\varepsilon : \Omega_n \longrightarrow R^+$ are continuous. Hence each $P : \Omega_n \longrightarrow cl(L^1(T,X))$ has the closed graph. An application of the Castaing & Jacobs Theorem once more yields the measurability of P. Now the Kuratowki & Ryll-Nardzewski Theorem gives a desired selection.   □

4. Carathéodory type selections

Combining the results concerning continuous and measurable selections of Aumann integrals we can obtain Carathéodory type selections. However arguments concerning continuous selections have to be used more thoroughly as in the proof of the Theorem 53. The reason is that the sets

$$\left\{ u \in K\left(\omega, s\right) : \left| a\left(\omega, s\right) - \int_T u\left(t\right) \mu\left(dt\right) \right| < \varepsilon\left(\omega, s\right) \right\}$$

are neither convex nor decomposable.

THEOREM 54. *Let* $K : \Omega \times S \longrightarrow dcl\left(T, X\right)$ *be a l.C. multifunction and assume that* $K\left(\omega, s\right)$ *is for each* $\left(\omega, s\right) \in \Omega \times S$ *integrably bounded by* $\beta_{\omega, s} \in L^1\left(T\right)$. *Fix a Carathéodory type selection*

$$a\left(\omega, s\right) \in cl \int_T K\left(\omega, s\right).$$

Then for every Carathéodory function $\varepsilon : \Omega \times S \longrightarrow R^+$ *one can find a Carathéodory selection* $k\left(\omega, s\right) \in K\left(\omega, s\right)$ *such that for every* $\left(\omega, s\right) \in \Omega \times S$ *the inequality*

$$\left| a\left(\omega, s\right) - \int_T k\left(\omega, s\right) \mu\left(dt\right) \right| < \varepsilon\left(\omega, s\right)$$

holds.

PROOF. We shall use an "extension method". By the Theorem 30 we may find in Ω an increasing family of compact subsets $\{\Omega_n\}_{n \in N}$ with $\mathsf{m}\left(\Omega \backslash \Omega_n\right) \underset{n \longrightarrow \infty}{\longrightarrow} 0$ such that each $K : \Omega_n \times S \longrightarrow dcl\left(T, X\right)$ is l.s.c., while $a : \Omega_n \times S \longrightarrow X$ and $\varepsilon : \Omega_n \times S \longrightarrow R^+$ are continuous.

Starting with the continuous selection $a : \Omega_1 \times S \longrightarrow X$ of $\mathcal{A} : \Omega_1 \times S \longrightarrow cc\left(X\right)$ we choose, by the Theorem 50, a continuous selection $k : \Omega_1 \times S \longrightarrow L^1\left(T, X\right)$ such that for $\left(\omega, s\right) \in \Omega_1 \times S$ the inequality

$$\left| a\left(\omega, s\right) - \int_T k\left(\omega, s\right)\left(t\right) \mu\left(dt\right) \right| < \frac{\varepsilon\left(\omega, s\right)}{2}$$

holds. Define $K_1 : \Omega \times S \longrightarrow dcl\,(T, X)$ by the formula

$$K_1\,(\omega, s) = \begin{cases} \{k\,(\omega, s)\} & for \quad \omega \in \Omega_1 \\ K\,(\omega, s) & for \quad \omega \notin \Omega_1 \end{cases}$$

and notice that each continuous selection of K_1 is a continuous extension of k. By construction the mapping $K_1 : \Omega_2 \times S \longrightarrow dcl\,(T, X)$ is $l.s.c.$ and so is for it's Aumann integral $\mathcal{A}_1\,(\omega, s) = cl \int_T K_1\,(\omega, s)$. Moreover for all $(\omega, s) \in \Omega \times S$ we have

$$d\,(a\,(\omega, s)\,, \mathcal{A}_1\,(\omega, s)) < \frac{\varepsilon\,(\omega, s)}{2}$$

By the Proposition 35 there exists a continuous selection $a_1 : \Omega_2 \times S \longrightarrow X$ of $\mathcal{A}_1 : \Omega_2 \times S \longrightarrow cc\,(X)$ such that

$$|a\,(\omega, s) - a_1\,(\omega, s)| < \frac{\varepsilon\,(\omega, s)}{2} + \frac{\varepsilon\,(\omega, s)}{2^3}.$$

The function k can be therefore continuously extended to a selection, denoted again by k, of $K : \Omega_2 \times S \longrightarrow dcl\,(T, X)$ in such way that for all $(\omega, s) \in \Omega_2 \times S$ we have

$$\left| a_1\,(\omega, s) - \int_T k\,(\omega, s)\,(t)\,\mu\,(dt) \right| < \frac{\varepsilon\,(\omega, s)}{2^3}.$$

Thus for $(\omega, s) \in \Omega_2 \times S$ an estimate

$$\left| a\,(\omega, s) - \int_T k\,(\omega, s)\,(t)\,\mu\,(dt) \right| < \frac{\varepsilon\,(\omega, s)}{2} + \frac{\varepsilon\,(\omega, s)}{2^2}$$

holds. Having already extended k to a continuous selection of $K : \Omega_n \times S \longrightarrow dcl\,(T, X)$ such that for $(\omega, s) \in \Omega_n \times S$

$$\left| a\,(\omega, s) - \int_T k\,(\omega, s)\,(t)\,\mu\,(dt) \right| <$$

$$< \frac{\varepsilon\,(\omega, s)}{2} + \frac{\varepsilon\,(\omega, s)}{2^2} + ... + \frac{\varepsilon\,(\omega, s)}{2^n}$$

we can define $K_n : \Omega \times S \longrightarrow dcl\,(T, X)$ by the formula

$$K_n\,(\omega, s) = \begin{cases} \{k\,(\omega, s)\} & for \quad \omega \in \Omega_n \times S \\ K\,(\omega) & for \quad \omega \notin \Omega_n \times S \end{cases}.$$

The mapping $K_n : \Omega_{n+1} \times S \longrightarrow dcl\,(T, X)$ is *l.s.c.* and the same holds for it's Aumann integral

$$\mathcal{A}_n\,(\omega, s) = cl \int\limits_T K_n\,(\omega, s)\,.$$

Moreover, for all $(\omega, s) \in \Omega \times S$ we have

$$d\,(a\,(\omega, s)\,, \mathcal{A}_n\,(\omega, s)) < \frac{\varepsilon\,(\omega, s)}{2} + \frac{\varepsilon\,(\omega, s)}{2^2} + ... + \frac{\varepsilon\,(\omega, s)}{2^n}.$$

Applying the Proposition 35 once more we get a continuous selection $a_n : \Omega_{n+1} \times S \longrightarrow X$ of $\mathcal{A} : \Omega_{n+1} \times S \longrightarrow cc\,(X)$ such that

$$|a\,(\omega, s) - a_n\,(\omega, s)| <$$

$$< \frac{\varepsilon\,(\omega, s)}{2} + \frac{\varepsilon\,(\omega, s)}{2^2} + ... + \frac{\varepsilon\,(\omega, s)}{2^n} + \frac{\varepsilon\,(\omega, s)}{2^{n+2}}.$$

Now the function k can be continuously extended to a selection, still denoted by k, of $K_n : \Omega_{n+1} \times S \longrightarrow dcl\,(T, X)$ having property that for all $(\omega, s) \in \Omega_{n+1} \times S$

$$\left| a_n\,(\omega, s) - \int\limits_T k\,(\omega, s)\,(t)\,\mu\,(dt) \right| < \frac{\varepsilon\,(\omega, s)}{2^{n+2}}.$$

Thus

$$\left| a\,(\omega, s) - \int\limits_T k\,(\omega, s)\,(t)\,\mu\,(dt) \right| \leq$$

$$\leq \frac{\varepsilon\,(\omega, s)}{2} + \frac{\varepsilon\,(\omega, s)}{2^2} + ... + \frac{\varepsilon\,(\omega, s)}{2^n} + \frac{\varepsilon\,(\omega, s)}{2^{n+1}}.$$

Continuing the described above procedure we finally get a Carathéodory type extension of k on $\Omega_n \times S$ such that for all $(\omega, s) \in \Omega_{n+1} \times S$ the inequality

$$\left| a\,(\omega, s) - \int\limits_T k\,(\omega, s)\,(t)\,\mu\,(dt) \right| <$$

$$< \frac{\varepsilon\,(\omega, s)}{2} + \frac{\varepsilon\,(\omega, s)}{2^2} + ... + \frac{\varepsilon\,(\omega, s)}{2^n} + ... = \varepsilon\,(\omega, s)$$

is valid, what ends the proof. $\square$

CHAPTER 13

Fixed points for multivalued contractions

Recall that $P : T \longrightarrow b(X)$ is called a Lipschitz multivalued mapping iff there is an $\alpha \geq 0$ such that for all $t, z \in T$ the following condition

(13.1) $$d_H(P(t), P(z)) \leq \alpha d(t, z)$$

holds. If $\alpha \in [0, 1)$ then P is called a contraction. A selection $p : T \longrightarrow X$ of P which is a Lipschitz function we call a Lipschitz selection. If selection p is a contraction then we say it is a contractive selection.

We say that $t \in T \subset X$ is a fixed point of $P : T \longrightarrow N(X)$ iff $t \in P(t)$. The set of fixed points we shall denote by $\mathcal{F}ix(P)$, i.e.

$$\mathcal{F}ix(P) = \{t : t \in P(t)\}$$

We also say that a subset $A \subset X$ has fixed point property iff each compact mapping $p : A \longrightarrow A$ admits a fixed point.

Problems of existence of fixed points and their nature have been attracted many generations of mathematicians. In the bunch of the results concerning this subject special role plays the famous Banach Contraction Principle saying that any pointwise contraction $p : X \longrightarrow X$ admits exactly one fixed point which can be achieved as the limit of a sequence of consequtive iterations. This fact has plenty of applications in various areas of mathematics, in particular in ordinary and partial differential equations. Since our book is motivated by differential inclusions we need a version of the Banach Contraction Principle in the multivalued case. Such problems for multivalued contractions $P : X \longrightarrow N(X)$ one could try possibly to attack by the contractive selections. However in general it is impossible. In euclidean spaces the only known results concern the existence of Lipschitz selections for Lipschitz multivalued mappings $P : R^l \longrightarrow clco(R^l)$. But if P is Lipschitz with constant α then there are Lipschitz selections with constant $\alpha\sqrt{n}$ and may not be ones with smaller constant. In particalur it means that multivalued contractions may not have contractive selections. The Lipschitz selections may be applicable for differential inclusions in R^l since in the space $L^1(I, R^l)$ we shall work exclusively with equivalent *Bielecki norms*. In infinitely dimensional Banach spaces there are

Lipschitz multivalued mappings $P : X \longrightarrow clco\,(X)$ which admits no Lipschitz selections. The first example of this type belongs to Yost [**220**]. He has also shown that a problem of nonexistence of Lipschitz selections is generic for finitely dimensional Banach spaces. Namely, we have

THEOREM 55 (Yost). *There exists a Lipschitz mapping from $(b\,(X)\,, d_H)$ to X if and only if X is finitely dimensional Banach space.*

Problem of the existence of Lipschitz selections has also been examined by Przesławski&Yost [**201**], Valadier [**217**], Łojasiewicz [**151**], Bressan [**37**] , Ornelas [**189**], Cellina&Ornelas [**49**], Dentscheva [**58**], [**59**] and many others. Main effort was directed towards the convex-valued multifunctions and in most of the papers it were exploited Steiner points in various settings or so called metric projections. There are also known special kinds of the Castaing representations for Lipschitz multivalued mappings $P : X \longrightarrow clco\,(R^l)$. However an approach via Lipschitz selections does not fit to mappings with decomposable values since this kind of results is simply unknown. An existence of fixed points and properties of the set $\mathcal{F}ix\,(P)$ we obtain by an use of continuous selections. We also consider a dependence on a parameter s of the sets $\mathcal{F}ix\,(P\,(s))$ for multivalued mappings $P : S \times X \longrightarrow N\,(X)$.

In what follows we impose the following properties for $P : S \times X \longrightarrow N\,(X)$:

CONDITION 1. $P\,(\cdot, x)$ *is continuous for every x*;

CONDITION 2. *there is a continuous $\alpha : s \longrightarrow (0,1)$ such that $P\,(s, \cdot)$ is a contraction with the constant $\alpha\,(s)$.*

1. Convex case

Let $P : S \times X \longrightarrow clco\,(X)$. Denote by $\mathcal{F}\,(s) = \mathcal{F}ix\,\{P\,(s)\}$ the set of fixed points of $P\,(s, \cdot)$. We begin with the following

THEOREM 56. *Let $P : S \times X \longrightarrow clco\,(X)$ be a multivalued mapping satisfying Conditions 1 and 2. Then, for every $s \in S$, the set $\mathcal{F}\,(s)$ is an absolute retract. Moreover, there exists a continuous mapping $r : S \times X \longrightarrow X$ which for every $s \in S$ establishes the retraction of X onto $\mathcal{F}\,(s)$.*

PROOF. Denote by

$$\varphi\,(s, x) = d\,(x, P\,(s, x))$$

and notice that, by the Condition 2, the mapping φ is Lipschitz in x and continuous in s. Hence it is $u.s.c.$ in (s, x). Moreover

$$x \in \mathcal{F}(s) \quad iff \quad \varphi(s, x) = 0.$$

Take

$$T = (S \times X) \setminus gr\mathcal{F} = \{(s, x) : x \notin \mathcal{F}(s)\} = \{(s, x) : \varphi(s, x) > 0\}$$

and notice that T is open.

For every $(s, x) \in T$ define

$$R(s, x) = cl\left\{ z \in P(s, x) : |z - x| < \frac{1 + \alpha(s)}{2\alpha(s)}\varphi(s, x) \right\}.$$

Employing the Proposition 58e) we see that the multivalued mapping $R : T \longrightarrow clco(X)$ is $l.s.c.$ and therefore by the Michael Selection Theorem it admits a continuous selection p. Extend p on $S \times X$ by setting

$$p(s, x) = x \quad for \quad x \in \mathcal{F}(s).$$

By construction for all $(s, x) \in S \times X$ we have

(13.2) $$|p(s, x) - x| \leq \frac{1 + \alpha(s)}{2\alpha(s)}\varphi(s, x)$$

and

$$p(s, x) \in P(s, x).$$

We claim that $p : S \times X \longrightarrow X$ is continuous. Clearly, it is enough to check it for $(s, x) \in gr(\mathcal{F})$. Fix $x \in \mathcal{F}(s)$ and take arbitrary $(s_n, x_n) \longrightarrow (s, x)$. Then by continuity

$$\lim_{n \to \infty} \varphi(s_n, x_n) = \lim_{n \to \infty} d(x_n, P(s_n, x_n)) = d(x, P(s, x)) = 0.$$

Now from (13.2) we obtain

$$p(s_n, x_n) \longrightarrow x = p(s, x),$$

proving our claim.

Set

$$r_1(s, x) = p(s, x)$$

and inductively

$$r_{n+1}(s, x) = p(s, r_n(s, x)).$$

Obviously each $r_n(s, x)$ is continuous with

(13.3) $$r_{n+1}(s, x) \in P(s, r_n(s, x))$$

and

(13.4) $$|r_{n+1}(s, x) - r_n(s, x)| \leq \frac{1 + \alpha(s)}{2\alpha(s)}\varphi(s, r_n(s, x)).$$

We shall show that r_n converges locally uniformly to a continuous function r which occures to be a required retracion. Indeed, from (13.3) and (13.4) we see that for all $(s, x) \in S \times X$

$$|r_{n+1}(s, x) - r_n(s, x)| \leq \frac{1 + \alpha(s)}{2\alpha(s)} d(r_n(s, x), P(s, r_n(x))) \leq$$

$$\leq \frac{1 + \alpha(s)}{2\alpha(s)} d(P(s, r_{n-1}(x)), P(s, r_n(x))) \leq$$

$$\leq \frac{1 + \alpha(s)}{2} |r_{n-1}(s, x) - r_n(s, x)|.$$

Therefore

$$|r_{n+1}(s, x) - r_n(s, x)| \leq \left(\frac{1 + \alpha(s)}{2}\right)^n \varphi(s, x).$$

Since the right-hand side of the above inequality is locally bounded then r_n converges locally uniformly. Therefore the mapping r given by

$$r(s, x) = \lim_{n \to \infty} r_n(s, x)$$

is continuous on $S \times X$. Moreover, for $x \in \mathcal{F}(s)$ we have

$$r(s, x) = x.$$

Passing to the limit in (13.3) we obtain

$$r(s, x) \in P(s, r(s, x)),$$

what shows that for every $s \in S$

$$r(s, \cdot) : X \longrightarrow \mathcal{F}(s).$$

This ends the proof.   □

COROLLARY 13 (Ricceri). *Let $P : X \longrightarrow clco(X)$ be a contraction. Then P admits a fixed point. Moreover, the set of fixed points*

$$\mathcal{F}ix(P) = \{x : x \in P(x)\}$$

is an absolute retract.

2. Decomposable case

Problem of existence of fixed points and their properties make also sense for multivalued contractions with decomposable values. Roughly speaking obtained results are analogous to the convex case, however again methods used have to be adopted to the present situation. Let $\mathcal{K} : L^p(T, X) \longrightarrow dcl_p(T, X)$ be an $\alpha - contraction$ and denote by

$$\varphi(u)(t) = ess\inf\{|u(t) - v(t)| : v \in \mathcal{K}(u)\}.$$

LEMMA 8. *The mapping $\varphi : L^p(T, X) \longrightarrow L^p(T, R)$ is Lipschitz with constant $2\alpha + 1$.*

PROOF. Fix $\varepsilon > 0$ and take any $u, v \in L^p(T, X)$. From the Proposition 54 there exist $w \in K(u)$ and $z \in K(v)$ fulfilling inequalities

$$|u(t) - w(t)| \le \varphi(u)(t) + \varepsilon \quad a.e. \ in \ T$$

and

$$|v(t) - z(t)| \le \varphi(v)(t) + \varepsilon \quad a.e. \ in \ T.$$

Moreover, there are $a \in K(u)$ and $b \in K(v)$ such that

(13.5) $$\|a - z\|_p \le d(z, K(u))\|_p + \varepsilon \le \alpha\|u - v\|_p + \varepsilon$$

and

(13.6) $$\|b - w\|_p \le d(w, K(v))\|_p + \varepsilon \le \alpha\|u - v\|_p + \varepsilon.$$

Then *a.e. in T* we have

$$\varphi(v)(t) - \varphi(u)(t) \le |v(t) - b(t)| - |u(t) - w(t)| + \varepsilon \le$$

$$\le |v(t) - u(t)| + |u(t) - b(t)| - |u(t) - w(t)| + \varepsilon \le$$

$$\le |v(t) - u(t)| + |b(t) - w(t)| + |a(t) - z(t)| + \varepsilon.$$

Similarly

$$\varphi(u)(t) - \varphi(v)(t) \le |v(t) - u(t)| + |b(t) - w(t)| + |a(t) - z(t)| + \varepsilon$$

and therefore

$$\|\varphi(u) - \varphi(v)\|_p \le \|u - v\|_p + \|w - b\|_p + \|a - z\|_p + \varepsilon.$$

Employing (13.5) and (13.6) we get

$$\|\varphi(u) - \varphi(v)\|_p \le (2\alpha + 1)\|u - v\|_p + 3\varepsilon.$$

But $\varepsilon > 0$ is arbitrary, so the latter shows our claim.   □

From now on let $K : S \times L^p(T, X) \longrightarrow dcl_p(T, X)$ be a multivalued mapping with the following properties:

CONDITION 3. *$K(\cdot, u)$ is continuous for every u;*

CONDITION 4. *there is a continuous $\alpha : s \longrightarrow (0, 1)$ such that $K(s, \cdot)$ is a contraction with a constant $\alpha(s)$.*

Denote by $\mathcal{F}(s)$ the set of fixed points of $K(s, \cdot)$ i.e.

$$\mathcal{F}(s) = \{u : u \in K(s, u)\}.$$

THEOREM 57 (Bressan-Cellina-Fryszkowski [**36**]). *Let* $\mathcal{K} : S \times L^p(T, X) \longrightarrow$ $dcl_p(T, X)$, $1 \leq p < \infty$, *be a multifunction satisfying Conditions 3 and 4. Then each* $\mathcal{F}(s)$ *is a retract of* $L^p(T, X)$. *Moreover, it can be found a continuous mapping* $r : S \times L^p(T, X) \longrightarrow L^p(T, X)$ *which for every* $s \in S$ *establishes the retraction of* $L^p(T, X)$ *onto* $\mathcal{F}(s)$.

PROOF. Denote by

$$\varphi(s, u)(t) = ess\inf\{|u(t) - v(t)| : v \in \mathcal{K}(s, u)\}$$

and notice that, by the Lemma 8 and Condition 3, the mapping $\varphi :$ $S \times L^p(T, X) \longrightarrow L^p(T, R)$ is Lipschitz in u and continuous in s. Hence it is continuous in (s, u). Moreover

$$\|\varphi(s, u)\|_p = d_p(u, \mathcal{K}(s, u)).$$

One can observe that

$$u \in \mathcal{F}(s) \quad iff \quad \varphi(s, u) = 0.$$

In further steps we mimic our proof of the Theorem 56.

Denote by

$$W = \{S \times L^p(T, X)\} \setminus gr\mathcal{F} = \{(s, u) : u \notin \mathcal{F}(s)\} =$$
$$= \left\{(s, u) : \|\varphi(s, u)\|_p > 0\right\}$$

and notice that W is open.

For every $(s, u) \in W$ define

$$\psi(s, u) = \frac{(1 + \alpha(s))\left(\|\varphi(s, u)\|_p + \varphi(s, u)\right)}{4\alpha(s)}$$

and observe that

(13.7) $$\|\psi(s, u)\|_p \leq \frac{1 + \alpha(s)}{2\alpha(s)}\|\varphi(s, u)\|_p.$$

Consider the sets

$$R(s, u) = cl\{z \in \mathcal{K}(s, u) : |z(t) - u(t)| < \psi(s, u)(t) \quad a.e. \ in \ T\}.$$

Observe that by the Proposition 54 each $R(s, u)$ is a nonempty decomposable set. Hence using the Proposition 58 we can establish the l.s.c. of the multivalued mapping $R : W \longrightarrow dcl_p(T, X)$. So, by the Bressan-Colombo-Fryszkowski Theorem, it admits a continuous selection $k : W \longrightarrow L^p(T, X)$. Extend k on $S \times L^p(T, X)$ by setting

$$k(s, u) = u \quad for \quad u \in \mathcal{F}(s).$$

By construction we have

$$k(s, u) \in \mathcal{K}(s, u)$$

and

$$|k(s,u)(t) - u(t)| \leq \psi(s,u) \quad a.e. \ in \ T.$$

The latter together with (13.7) shows that

(13.8) $$\|k(s,u) - u\|_p \leq \frac{1+\alpha(s)}{2\alpha(s)} \|\varphi(s,u)\|_p$$

We claim that $k : S \times L^p(T,X) \longrightarrow L^p(T,X)$ is continuous. Clearly, it is enough to verify this for $(s,u) \in gr(\mathcal{F})$. Fix $u \in \mathcal{F}(s)$ and take arbitrary $(s_n, u_n) \longrightarrow (s,u)$. Then by continuity

$$\lim_{n \to \infty} \|\varphi(s_n, u_n)\|_p = \lim_{n \to \infty} d_p(u_n, \mathcal{K}(s_n, u_n)) = d_p(u, \mathcal{K}(s,u)) = 0.$$

Therefore from (13.2) we obtain

$$k(s_n, u_n) \longrightarrow u = k(s,u),$$

ensuring our claim.

Further we shall apply an iteration technique. Set

$$r_1(s,u) = k(s,u)$$

and, indictively,

$$r_{n+1}(s,u) = k(s, r_n(s,u)).$$

Obviously each $r_n(s,u)$ is continuous and

(13.9) $$r_{n+1}(s,u) \in \mathcal{K}(s, r_n(s,u)).$$

We shall show that r_n converges locally uniformly to a continuous function r which is a required retraction. Indeed, from (13.2) and (13.8) we see that for $n = 1, 2, ...$

$$\|r_{n+1}(s,u) - r_n(s,u)\|_p \leq \frac{1+\alpha(s)}{2\alpha(s)} \|\varphi(s, r_n(s,u))\|_p.$$

Therefore

$$\|r_{n+1}(s,u) - r_n(s,u)\|_p \leq \frac{1+\alpha(s)}{2\alpha(s)} d_p(r_n(s,u), \mathcal{K}(s, r_n(s,u))) \leq$$

$$\leq \frac{1+\alpha(s)}{2\alpha(s)} d_H(\mathcal{K}(s, r_{n-1}(u)), \mathcal{K}(s, r_n(s,u))) \leq$$

$$\leq \frac{1+\alpha(s)}{2} \|r_n(s,u) - r_{n-1}(s,u)\|_p$$

and hence

$$\|r_{n+1}(s,u) - r_n(s,u)\|_p \leq \left(\frac{1+\alpha(s)}{2}\right)^{n+1} \|\varphi(s,u)\|_p.$$

Since the right-hand side of the above inequality is locally bounded, so r_n converges locally uniformly. Thus the mapping

$$r(s, u) = \lim_{n \to \infty} r_n(s, u)$$

is continuous. Moreover, for $u \in \mathcal{F}(s)$ we have

$$r(s, u) = u.$$

Passing to the limit in (13.9) we obtain

$$r(s, u) \in \mathcal{K}(s, r(s, u)),$$

what shows that for every $s \in S$ we have $\mathcal{F}(s) \neq \emptyset$ and

$$r(s, \cdot) : L^p(T, X) \longrightarrow \mathcal{F}(s).$$

This ends the proof.   □

Since for given $K \in dcl_p(T, X)$ the constant multivalued mapping $\mathcal{K}(u) = K$ is a contraction with the set of fixed points coinciding with K then as a conclusion from the previous theorem we have

THEOREM 58 (Bressan & Colombo [**35**]). *Any decomposable set* $K \in dcl_p(T, X)$, $p < \infty$, *is a retract of* $L^p(T, X)$.

COROLLARY 14 (Cellina [**44**], Fryszkowski [**75**]). *Any decomposable set* $K \in dcl_p(T, X)$, $p < \infty$, *possess the fixed point property.*

PROOF. Fix $K \in dcl_p(T, X)$ and take a retraction $r : L^p(T, X) \longrightarrow K$. Consider a compact mapping $\phi : K \longrightarrow K$ and set $S = clco\phi(K)$. Observe that S is compact and the mapping $\phi \circ r : L^p(T, X) \longrightarrow S$ is continuous. Employing the Schauder Theorem one can easily conclude that the mapping $\phi \circ r$ restricted to S admits a fixed point s which has to be in $\phi(K) \subset K$. Hence $r(s) = s$ and therefore s is also a fixed point of ϕ.   □

CHAPTER 14

Operator and differential inclusions

Let X and Y be Banach spaces and let $T \subset R^k$ be a domain. Consider a continuous linear operator

$$A : D(A) \subset L^p(T, X) \longrightarrow L^q(T, Y).$$

We shall assume that A admits the right inverse, i.e. there exists a continuous linear bounded operator $R : L^q(T, Y) \longrightarrow L^p(T, X)$, such that

$$AR = I.$$

By an operator inclusion we wean a relation

(14.1) $$Au \in F(t, u),$$

where $F : T \times X \longrightarrow N(Y)$ is a given multivalued mapping.

A solution of (14.1) can be understand in many ways. We shall point out just two of them:

classical solution - that is a function $u \in D(A)$ such that

$$(Au)(t) \in F(t, u(t)) \quad a.e. \text{ in } T;$$

mild (or absolutely continuous solution) - that is a function $u = Rv$ such that

$$v(t) \in F(t, (Rv)(t)) \quad a.e. \text{ in } T.$$

Together with (14.1) we can also consider so called "relaxed" operator inclusions

(14.2) $$Au \in clcoF(t, u).$$

For the operator and differential inclusions the main effort of mathematicians has been concentrated on existence, properties of solutions and connections of solutions to the problem (14.2) with (14.1). In this book we want to give a survey for this kind of problems just for mild solutions. We start our considerations with an observation that for a function $u = Rv$ a property of being a mild solution to (14.1) is equivalent to the fact that v is a fixed point of the multifunction $K : L^p(T, X) \longrightarrow N(L^q(T, Y))$ defined by

$$K(v) = \{w \in L^p(T, X) : w(t) \in F(t, (Rv)(t)) \quad a.e. \text{ in } T\}.$$

Therefore the most consistent examination of properties of (mild) solutions to (14.1) can be concluded from corresponding facts for the set $\mathcal{F}ix\,(K)$.

As a model form of operator inclusions the reader could have in mind so called differential inclusions

$$(14.3) \qquad \begin{cases} u' \in F\,(t,u) \\ u\,(0) = \zeta \end{cases}$$

where $t \in I = [0,1]$. In this case the right inverse to the operator $A = \frac{d}{dt}$ is an operator $R = \mathcal{I}$ given by

$$u\,(t) = \mathcal{I}\,(\zeta,v)\,(t) = \zeta + \int\limits_0^t v\,(\tau)\,d\tau.$$

By a solution (mild or absolutely continuous solution of (14.3) we mean a function $u = \mathcal{I}\,(\zeta,v)$ determined by $v \in L^1\,(I,X)$ such that

$$v\,(t) \in F\,(t,u\,(t)) = F\,(t,\mathcal{I}\,(\zeta,v)\,(t)) \quad a.e.\ in\ I.$$

In other words, the function $u = \mathcal{I}\,(\zeta,v)$ is a mild solution of (14.3) if v is a fixed point of the multivalued mapping $K\,(v)$ given by

$$K\,(v) = \left\{ w \in L^1\,(I,X) : w\,(t) \in F\,(t,\mathcal{I}\,(\zeta,v)\,(t)) \quad a.e.\ in\ I \right\}.$$

By the solution set (the set of all trajectories) to (14.3) it is usually meant a collection

$$\mathcal{R} = \left\{ u = \mathcal{I}\,(\zeta,v) : v \in \mathcal{F}ix\,(K) \right\}.$$

We should also point out that most of the properties of the solution sets are the consequences of properties of the set of fixed points, since $\mathcal{R}$ is a linear transformation of $\mathcal{F}ix\,(K)$ by $\mathcal{I}$.

1. Filippov Lemma

We shall also consider parametrized differential inclusions

$$(14.4) \qquad \begin{cases} u' \in F\,(t,u,s) \\ u\,(0) = \zeta\,(s) \end{cases}$$

with a multifunction $F : I \times X \times S \longrightarrow cl\,(X)$. About F we shall impose the following properties:

  i. F is $\mathcal{L} \otimes \mathcal{B}\,(X \times S) - measurable$ in (t,x,s);
  ii. for any (t,x) the mapping $F\,(t,x,\cdot)$ is $l.s.c.$;
  iii. there exists a continuous mapping $l : S \longrightarrow L^1\,(I,R^+)$ such that for every $x_1, x_2 \in X$ and any $s \in S$ the inequality

$$d_H\,(F\,(t,x_1,s)\,,F\,(t,x_2,s)) \le l\,(s)\,(t)\,|x_1 - x_2| \quad a.e.\ in\ I$$

holds;

iv. for any continuous mappings $(\zeta, z) : S \longrightarrow X \times L^1(I, X)$ there exists a continuous mapping $\beta = \beta_{\zeta, z} : S \longrightarrow L^1(I, R)$ such that for every $s \in S$ we have

$$d_H(z(s)(t), F(t, \mathcal{I}(\zeta(s), z(s))(t))) \le \beta(s)(t) \quad a.e. \ in \ I.$$

Denote by $\mathcal{R}(s)$ the solution set to (14.4) We are interested in the properties of a mapping $\mathcal{R}(\cdot)$, especially whether the sets $\mathcal{R}(s)$ are nonempty and if so, do $\mathcal{R}(\cdot)$ admits the continuous selection property. Before we go to some results for this kind of problems we have to notice that due to *ii.* and the Proposition 58 the assumption *iv.* may be replaced by an equivalent condition:

$iv_0.$ there exists a continuous mapping $\beta_0 : S \longrightarrow L^1(I, R)$ such that for every $s \in S$ we have

$$d_H(0, F(t, 0, s)) \le \beta_0(s)(t) \quad a.e. \ in \ I.$$

Indeed, it can be easily concluded from an inequality

$$d_H(z(s)(t), F(t, \mathcal{I}(\zeta(s), z(s))(t))) \le$$

$$\le |z(s)(t)| + d_H(0, F(t, 0, s)) + l(s)(t)|\mathcal{I}(\zeta(s), z(s))(t)| \quad a.e. \ in \ I.$$

To give a formulation of a continuous version of the Fillipov Lemma we put

$$m(s)(t) = \int_0^t l(s)(\tau) \, d\tau$$

and notice that $M : S \longrightarrow R^+$ given by $M(s) = \exp(m(s)(1))$ is continuous.

THEOREM 59 (Filippov Lemma [68]). *Assume that* $F : I \times X \times S \longrightarrow cl(X)$ *satisfies conditions i. − iii and* $iv_0.$ *and consider continuous mappings* $z : S \longrightarrow AC(I, X)$ *and* $\beta : S \longrightarrow L^1(T, R)$ *fulfilling for every* $s \in S$ *the inequality*

$$d_H(z'(s)(t), F(t, z(s)(t), s)) \le \beta(s)(t) \quad a.e. \ in \ I.$$

Then for every continuous functions $\zeta : S \longrightarrow X$ *and* $\varepsilon : S \longrightarrow R^+$ *there exists a continuous mapping* $u : S \longrightarrow L^1(I, X)$ *defining for every* $s \in S$ *a mild solution* $v(s) = \mathcal{I}(\zeta(s), u(s)) = \zeta(s) + \int_0^t u(s)(\tau) \, d\tau$ *of (14.4) such that*
  a)

$$|z'(s)(t) - v'(s)(t)| \le$$

$$\le \varepsilon(s) + l(s)(t) \exp\{m(s)(t)\} + l(s)(t)|z(s)(0) - \zeta(s)| +$$

$$+l\left(s\right)\left(t\right)\int_{0}^{t}\beta\left(s\right)\left(\tau\right)\exp\left\{m\left(s\right)\left(t\right)-m\left(s\right)\left(\tau\right)\right\}d\tau+\beta\left(s\right)\left(t\right)\quad a.e.\ in\ I;$$

and
b)

$$\left|z\left(s\right)\left(t\right)-v\left(s\right)\left(t\right)-\left(z\left(s\right)\left(0\right)-\zeta\left(s\right)\right)\right|\leq$$
$$\leq\left\{\varepsilon\left(s\right)+\left|z\left(s\right)\left(0\right)-\zeta\left(s\right)\right|\right\}\exp\left\{m\left(s\right)\left(t\right)\right\}+$$
$$+\int_{0}^{t}\beta\left(s\right)\left(\tau\right)\exp\left\{m\left(s\right)\left(t\right)-m\left(s\right)\left(\tau\right)\right\}d\tau.$$

PROOF. First of all observe that we may reduce the problem to the case

$$z\left(s\right)\equiv0\quad and\quad\zeta\left(s\right)\equiv0.$$

Indeed. Denote by

$$G\left(t,x,s\right)=F\left(t,x+z\left(s\right)\left(t\right)-z\left(s\right)\left(0\right)+\zeta\left(s\right),s\right)-z'\left(s\right)\left(t\right)$$

and consider the problem

$$(14.5) \qquad \begin{cases} x'\in G\left(t,x,s\right) \\ \quad x\left(0\right)=0 \end{cases}.$$

Then the function

$$u\left(s\right)=x\left(s\right)+z\left(s\right)-z\left(s\right)\left(0\right)+\zeta\left(s\right)$$

is a desired solution of (14.4) whenever $x\left(s\right)$ is a solution of (14.5) satisfying a) and b) with

$$\widetilde{\beta}\left(s\right)=\beta\left(s\right)+l\left(s\right)\left|z\left(s\right)\left(0\right)-\zeta\left(s\right)\right|\geq d_{H}\left\{0,G\left(t,0,s\right)\right\}.$$

Fix $\varepsilon:S\longrightarrow R^{+}$, set $\varepsilon_{n}=\frac{n+1}{n+2}\varepsilon$ and put

$$\beta_{n}\left(s\right)\left(t\right)=\int_{0}^{t}\beta\left(s\right)\left(\tau\right)\frac{\left(m\left(s\right)\left(t\right)-m\left(s\right)\left(\tau\right)\right)^{n-1}}{\left(n-1\right)!}d\tau+\frac{\left(m\left(s\right)\left(t\right)\right)^{n-1}}{\left(n-1\right)!}\varepsilon_{n}\left(s\right).$$

Using the integration by parts and repeating the Filippov's calculation (see the Proposition 59) we obtain

$$\int_{0}^{t}l\left(s\right)\left(\tau\right)\beta_{n}\left(s\right)\left(\tau\right)=\int_{0}^{t}\beta\left(s\right)\left(\tau\right)\frac{\left(m\left(s\right)\left(t\right)-m\left(s\right)\left(\tau\right)\right)^{n}}{n!}d\tau+$$

$$+\frac{\left(m\left(s\right)\left(t\right)\right)^{n}}{n!}\varepsilon_{n}\left(s\right)<\beta_{n+1}\left(s\right)\left(t\right)\quad a.e.\ in\ I.$$

We shall construct a Cauchy sequence of successive approximations $u_n(s) \in AC(I, X)$, $n = 1, 2, \ldots$ with the following properties:

(14.6) $\qquad\qquad u_n : S \longrightarrow AC(I, X) \quad$ are continuous;

(14.7) $\qquad\qquad u'_{n+1}(s)(t) \in F(t, u_n(s)(t), s) \quad$ a.e. in I;

(14.8) $\qquad |u'_{n+1}(s)(t) - u'_n(s)(t)| \leq l(s)(t) \beta_n(s)(t) \quad$ a.e. in I,

where for simplicity $l(s) \beta_0(s)$ is understood as $\beta(s) + \varepsilon_0(s)$. Observe that from (14.8) we then have

(14.9) $\qquad |u_{n+1}(s)(t) - u_n(s)(t)| < \beta_{n+1}(s)(t) \quad$ a.e. in I.

Set $u_0(s) = 0$ and denote by

$$F_0(s) = \{v \in L^1(I, X) : v(t) \in F(t, u_0(s)(t), s) \quad \text{a.e. in } I\}.$$

Consider the multifunction G_0 defined by

$$G_0(s) = cl\{v \in F_0(s) : |v(t)| < \beta(s)(t) + \varepsilon_0(s) \quad \text{a.e. in } I\}.$$

Then the mapping $G_0 : S \longrightarrow dcl(X)$ is $l.s.c.$ and therefore by the Bressan-Colombo-Fryszkowski Theorem there exists a continuous mapping $g_0 : S \longrightarrow L^1(I, X)$ such that

$$g_0(s)(t) \in F(t, u_0(s)(t), s) \quad \text{a.e. in } I$$

and

$$|g_0(s)(t)| \leq \beta(s)(t) + \varepsilon_0(s) = l(s) \beta_0(s) \quad \text{a.e. in } I.$$

Define

$$u_1(s)(t) = \int_0^t g_0(s)(\tau) \, d\tau$$

and notice that

$$|u_1(s)(t) - u_0(s)(t)| \leq \int_0^t |g_0(s)(\tau)| \, d\tau <$$

$$< \int_0^t \beta(s)(\tau) \, d\tau + \varepsilon_0(s) = \beta_1(s)(t) \quad \text{a.e. in } I.$$

Suppose that we have defined functions $u_0, u_1, \ldots, u_n$ satisfying (14.6), (14.7) and (14.8). Then by the lipschitzness we have

$$d(u'_n(s)(t), F(t, u_n(s)(t), s)) \leq$$

$$d(F(t, u_{n-1}(s)(t), s), F(t, u_n(s)(t), s)) \leq$$

$$\leq l(s)(t) |u_n(s)(t) - u_{n-1}(s)(t)| \quad \text{a.e. in } I.$$

The latter and (14.9) yield

$$d\left(u'_n\left(s\right)\left(t\right), F\left(t, u_n\left(s\right)\left(t\right), s\right)\right) < l\left(s\right)\left(t\right)\beta_n\left(s\right)\left(t\right) \quad a.e.\ in\ I.$$

Denote

$$F_n\left(s\right) = \left\{v \in L^1\left(I, X\right) : v\left(t\right) \in F\left(t, u_n\left(s\right)\left(t\right), s\right) \quad a.e.\ in\ I\right\}$$

and consider sets

$$G_n\left(s\right) = cl\left\{v \in F_n\left(s\right) : \left|v\left(t\right) - u'_n\left(s\right)\left(t\right)\right| < l\left(s\right)\left(t\right)\beta_n\left(s\right)\left(t\right) \quad a.e.\ in\ I\right\}.$$

By construction the sets $G_n\left(s\right)$ are nonempty and decomposable, while in virtue of the Proposition 58 the multivalued mapping $G_n : S \longrightarrow dcl\left(X\right)$ is $l.s.c.$. Employing again the Bressan-Colombo-Fryszkowski Theorem we obtain the existence of a continuous mapping

$$g_n : S \longrightarrow L^1\left(I, X\right)$$

such that

$$\left|g_n\left(s\right)\left(t\right) - u'_n\left(s\right)\left(t\right)\right| \le l\left(s\right)\left(t\right)\beta_n\left(s\right)\left(t\right) \quad a.e.\ in\ I.$$

Define

$$u_{n+1}\left(s\right)\left(t\right) = \int_0^t g_n\left(s\right)\left(\tau\right) d\tau$$

and observe that u_{n+1} satisfies (14.6), (14.7) and (14.8).

From (14.8) and (14.9) we conclude that

$$\left\|u_{n+1}\left(s\right) - u_n\left(s\right)\right\|_{AC} \le \beta_{n+1}\left(s\right)\left(1\right).$$

But

$$\beta_{n+1}\left(s\right)\left(1\right) \le \int_0^1 \beta\left(s\right)\left(\tau\right)\frac{\left(\left\|l\left(s\right)\right\|_1\right)^n}{n!} d\tau + \frac{\left[m\left(s\right)\left(1\right)\right]^n}{n!}\varepsilon_n\left(s\right).$$

Therefore

$$(14.10) \qquad \left\|u_{n+1}\left(s\right) - u_n\left(s\right)\right\|_{AC} \le \frac{\left(\left\|l\left(s\right)\right\|_1\right)^n}{n!}\left[\left\|\beta\left(s\right)\right\|_1 + \varepsilon\left(s\right)\right],$$

since

$$m\left(s\right)\left(t\right) - m\left(s\right)\left(z\right) = \int_z^t l\left(s\right)\left(\tau\right) d\tau \le \left\|l\left(s\right)\right\|_1 = m\left(s\right)\left(1\right).$$

Notice that the functions $s \longrightarrow \left\|\beta\left(s\right)\right\|_1$ and $s \longrightarrow \left\|l\left(s\right)\right\|_1$ are continuous. Therefore (14.10) means that the sequence $\left\{u_n\left(s\right)\right\}$ satisfies the Cauchy condition locally uniformly in some neighbourhood of any $s \in S$. Thus the mapping $u\left(s\right) = \lim u_n\left(s\right)$ is continuous from S into

$AC(I, X)$. We claim that u is a desired solution. To see that u is a solution of (14.5) let us observe that, by the lipschitzness, we have

$$d\left(u'_{n+1}(s)(t), F(t, u(s)(t), s)\right) \leq l(s)(t)|u_n(s)(t) - u(s)(t)| \quad a.e. \text{ in } I.$$

We shall check that u satisfies conditions $a)$ and $b)$.

Adding (14.8) for all n we obtain

$$\left|u'_{n+1}(s)(t) - u'_1(s)(t)\right| \leq \beta(s)(t) + \sum_{i=1}^{n} \left|u'_{i+1}(s)(t) - u'_i(s)(t)\right| + \varepsilon_0(s) \leq$$

$$\leq \beta(s)(t) + l(s)(t) \int_0^t \beta(s)(\tau) \left[\sum_{i=1}^{n} \frac{(m(s)(t) - m(s)(\tau))^{i-1}}{(i-1)!}\right] d\tau +$$

$$+\varepsilon(s)\, l(s)(t) \left[\sum_{i=1}^{n} \frac{(m(s)(t))^{i-1}}{(i-1)!}\right] + \varepsilon(s).$$

Similarly, summing up (14.9), we end up with

$$|u_{n+1}(s)(t) - u_1(s)(t)| \leq \sum_{i=1}^{n} |u_{i+1}(s)(t) - u_i(s)(t)| + \varepsilon_0(s) \leq$$

$$\leq \int_0^t \beta(s)(\tau) \left[\sum_{i=0}^{n} \frac{(m(s)(t) - m(s)(\tau))^{i}}{i!}\right] d\tau +$$

$$+\varepsilon(s) \left[\sum_{i=0}^{n} \frac{(m(s)(t))^{i}}{i!}\right] + \varepsilon(s).$$

Passing to the limit in both above inequalities we obtain $a)$ and $b)$. This ends the proof. $\qquad\qquad\qquad\qquad\qquad\qquad\qquad\qquad\qquad\square$

2. Continuous selections of solution sets

Denote by $\mathcal{R}(s)$ the solution set to (14.4)

$$\begin{cases} u' \in F(t, u, s) \\ u(0) = \zeta(s) \end{cases}$$

and by $\mathcal{F}(s) = \{u' : u \in \mathcal{R}(s)\}$. Our version of the Filippov Lemma 59 can be interpreted in particular in the following way:

THEOREM 60. *Let* $F : I \times X \times S \longrightarrow cl(X)$ *be a multifunction satisfying hypotheses* $i. - iv.$ *Then for each* $s \in S$ *the problem (14.4) admits a solution, i.e. the sets* $\mathcal{R}(s)$ *and* $\mathcal{F}(s)$ *are nonempty. Further, both multifunctions* $\mathcal{R} : S \longrightarrow AC(I, X)$ *and* $\mathcal{F} : S \longrightarrow L^1(I, X)$ *admit a continuous selection property.*

Now we shall discuss some others properties of the multivalued mapping $\mathcal{R} : S \longrightarrow AC\,(I, X)$, while to $\mathcal{F} : S \longrightarrow L^1\,(I, X)$ is devoted next section.

THEOREM 61. *Fix s_0 and a solution $u_0 \in \mathcal{R}\,(s_0)$. Then $\mathcal{R}\,(s)$ admits a continuous selection $r : S \longrightarrow AC\,(I, X)$ such that*

$$r\,(s_0) = u_0.$$

PROOF. Similarly as in the beginning of the proof of the Fillipov Lemma we may assume that $u_0 = 0$. Hence

$$0 \in F\,(t, 0, s_0) \quad a.e. \ \ in \ \ I.$$

Let us consider a multifunction $\widetilde{F} : I \times X \times S \longrightarrow cl\,(X)$ given by

$$\widetilde{F}\,(t, x, s) = \begin{cases} F\,(t, x, s) & if \ \ s \neq s_0 \\ \{0\} & if \ \ s = s_0 \end{cases}.$$

Clearly $\widetilde{F}$ satisfies hypotheses $i.$, $ii.$ and $iii..$ We shall observe also that $iv.$ holds and moreover we can choose continuous

$$\widetilde{\beta} : S \longrightarrow L^1\,(I, X)$$

in such a way that

$$\widetilde{\beta}\,(s_0) = 0.$$

From the definition of $\widetilde{F}$ we see that

$$d\left(0, \widetilde{F}\,(t, 0, s)\right) = d\,(0, F\,(t, 0, s)).$$

Consider a mapping $P : S \longrightarrow dcl\,(I, X)$ given by

$$P\,(s) = cl\left\{v \in L^1\,(I, X) : d\left(0, \widetilde{F}\,(t, 0, s)\right) < v\,(t) \quad a.e. \ \ in \ \ I\right\}$$

and notice that it is $l.s.c..$ Moreover

$$0 \in P\,(s_0).$$

So it admits a continuous selection $\widetilde{\beta} : S \longrightarrow L^1\,(I, X)$ with

$$\widetilde{\beta}\,(s_0) = 0.$$

Repeating the same construction as in the proof of the Filippov Lemma we see that $0 \in G_n\,(s_0)$ for all n. Proceeding carefully we can always pick a continuous selection g_n of G_n in such way that

$$g_n\,(s_0) = 0.$$

Hence the sequence of succesive approximations $u_n\,(s)$ satisfies for all n the condition

$$u_n\,(s_0) = 0$$

and the same holds for $u = \lim u_n$. This completes the proof. $\qquad\square$

Using exactly the same technics as in the proof of Theorem 61 we can deduce

THEOREM 62. *Under assumptions of the Theorem 61 the multi-function*

$$\mathcal{R} : S \longrightarrow cl\,\{AC\,(I,X)\}$$

is l.s.c.. Moreover there exists a countable family $\{r_n\}$ of continuous "branches" of solutions to (14.4) such that

$$\mathcal{R}\,(s) = cl\,\{r_n\,(s) : n = 1,2,...\}\,.$$

If we denote by $\mathcal{F}\,(s) = \{u' : u \in \mathcal{R}\,(s)\}$ then both mappings $\mathcal{R} : S \longrightarrow cl\,\{AC\,(I,X)\}$ and $\mathcal{F} : S \longrightarrow cl\,\{L^1\,(I,X)\}$ are l.s.c. and admit continuous selections.

At the moment we have to make a remark that the last result is strongest known one concerning the existence of continuous selections in functional spaces.

3. Filippov-Ważewski Relaxation Theorem

The classical Filippov-Ważewski result states that solution set $\mathcal{R}_F$ of the Cauchy problem (14.3)

$$\begin{cases} u' \in F\,(t,u)\,, \\ \quad u\,(0) = \zeta \end{cases}$$

is dense in the solution set $\overline{\mathcal{R}} = cl\mathcal{R} = \mathcal{R}_{clcoF}$ of the so called "relaxed (convexified) problem"

$$(14.11) \qquad \begin{cases} u' \in clcoF\,(t,u) \\ \quad u\,(0) = \zeta \end{cases},$$

whenever $F : I \times R^k \longrightarrow cl\,(R^k)$ satisfies conditions $i. - iii$ and iv_0.

We are going to present a continuous version of that result. Our considerations are based on continuous selections of Aumann integrals and the continuous version of the Filippov Lemma. Consider a family of Cauchy problems (14.4)

$$\begin{cases} u' \in F\,(t,u,s) \\ \quad u\,(0) = \zeta\,(s) \end{cases},$$

where $F : I \times X \times S \longrightarrow cl\,(X)$ satisfies conditions $i. - iv_0$. and additionally

  v. there exists $p \in L^1\,(I,R)$ such that for every $(s,x) \in S \times X$

$$\sup\,\{|u| : u \in F\,(t,x,s)\} \leq p\,(t) \quad a.e.\ in\ I.$$

By the relaxed (convexified) problem for (14.4) it is usually meant the Cauchy problem

(14.12)
$$\begin{cases} u' \in clcoF\,(t,u,s) \\ \quad u\,(0) = \zeta\,(s) \end{cases},$$

Denote by $\overline{\mathcal{R}}\,(s)$ the solution set for (14.12). Then the classical Filippov-Ważewski Theorem says that $\overline{\mathcal{R}}\,(s)$ coincides with $cl\,\{\mathcal{R}\,(s)\}$, where the closure is taken in $C\,(I,X)$. Using the notations from previous section our continuous version of the Filippov-Ważewski Theorem (see [8], [116]) is the following:

THEOREM 63 (Fryszkowski and Rzeżuchowski [81]). *Assume that* $F : I \times X \times S \longrightarrow cl\,(X)$ *satisfies conditions i. − iii. and v. while* $\zeta :$ $S \longrightarrow X$ *is continuous. Then for each continuous selection* $\overline{r}\,(s) \in$ $\overline{\mathcal{R}}\,(s)$ *and every continuous* $\varepsilon : S \longrightarrow R^+$ *with* $\inf\limits_{s \in S}\varepsilon\,(s) = \varepsilon_0 > 0$ *there exists a continuous selection* $r\,(s) \in \mathcal{R}\,(s)$ *such that for every* $s \in S$ *we have an estimate*

$$\|\overline{r}\,(s) - r\,(s)\|_C \le \varepsilon\,(s)\,M\,(s).$$

PROOF. Fix a continuous selection $\overline{r}\,(s) \in \overline{\mathcal{R}}\,(s)$ and a continuous $\varepsilon : S \longrightarrow R^+$. As it is done in the most proofs of the Filippov-Ważewski Theorem we shall first construct a continuous $k : S \longrightarrow L^1\,(I,X)$ which for every $s \in S$ is pointwisely *a.e.* close enough to the sets $F\,(t,\overline{r}\,(s)\,(t)\,,s)$ and next apply the Filippov Lemma. We begin with partition the interval I into subintervals $I_n = [t_n, t_{n+1}]$, $n = 0,1,...,j$ in such a way that

(14.13)
$$\int\limits_{I_n} p\,(t)\,dt \le \frac{\varepsilon_0}{8}.$$

Notice that since $\overline{r}\,(s)$ are solutions of (14.12) therefore the functions

$$s \longrightarrow \overline{r}\,(s)\,(t_{n+1}) - \overline{r}\,(s)\,(t_n)$$

are continuous selections of

$$cl\int\limits_{I_n} clcoF\,(t,\overline{r}\,(s)\,(t)\,,s)\,dt = cl\int\limits_{I_n} F\,(t,\overline{r}\,(s)\,(t)\,,s)\,dt.$$

Moreover,

(14.14)
$$|\overline{r}\,(s)\,(t_{n+1}) - \overline{r}\,(s)\,(t_n)| \le \frac{\varepsilon_0}{8}.$$

Therefore the Theorem 50 applied to the *l.s.c.* mapping

$$K_n : S \longrightarrow dcl\,(I_n, X)$$

given by
$$K_n(s) = \{u \in L^1(I_n, X) : u(t) \in F(t, \bar{r}(s)(t), s) \quad a.e. \text{ in } I_n\},$$

$n = 0, 1, ..., j$, yields the existence continuous selections

$$k_n : S \longrightarrow L^1(I_n, X)$$

such that for all $s \in S$ and any $n = 0, 1, ..., j$ we have

(14.15)
$$\left| \bar{r}(s)(t_{n+1}) - \bar{r}(s)(t_n) - \int_{I_n} k_n(s)(t)\, dt \right| < \frac{\varepsilon(s)}{4j}.$$

Let $u : S \longrightarrow AC(I, X)$ be such mapping that

$$u(s)(0) = \zeta(s)$$

and

$$u'(s)(t) = k_n(s)(t) \quad a.e. \text{ in } I_n.$$

Hence $u(s)(t_{n+1}) - u(s)(t_n) = \int_{I_n} k_n(s)(t)\, dt$ and therefore by (14.15)

we have

$$|\bar{r}(s)(t_{n+1}) - \bar{r}(s)(t_n) - u(s)(t_{n+1}) - u(s)(t_n)| < \frac{\varepsilon(s)}{4j}.$$

Summing up these inequalities we conclude that for every $n = 0, 1, ..., j$

$$|\bar{r}(s)(t_n) - u(s)(t_n)| < \frac{\varepsilon(s)}{4}.$$

But both $\bar{r}'(s)(\cdot)$ and $u'(s)(\cdot)$ are integrably bounded by p. So by (14.13) we get that for every $s \in S$ and all $t \in T$ an estimate

(14.16)
$$|\bar{r}(s)(t) - u(s)(t)| < \frac{\varepsilon_0}{4} + \frac{\varepsilon(s)}{4} \le \frac{\varepsilon(s)}{2}.$$

holds. Since by construction

$$u'(s)(t) \in F(t, \bar{r}(s)(t), s) \quad a.e. \text{ in } I$$

then using $ii.$ we obtain

(14.17)
$$d(u'(s)(t), F(t, u(s)(t), s)) \le \frac{\varepsilon(s)\, l(s)(t)}{2} \quad a.e. \text{ in } I.$$

Now from the Filippov Lemma applied for $\beta(s) = \frac{\varepsilon(s)l(s)}{4}$ and $\varepsilon(s)$ replaced with $\frac{\varepsilon(s)}{2}$ it follows that there exists a continuous selection $r(s) \in \mathcal{R}(s)$ such that for all $(s, t) \in S \times I$ is

$$|r(s)(t) - u(s)(t)| \le$$

$$\leq \frac{\varepsilon\left(s\right)}{2}\left\{\exp\left\{m\left(s\right)\left(t\right)\right\} + \int_{I} l\left(s\right)\left(\tau\right)\exp\left\{m\left(s\right)\left(t\right) - m\left(s\right)\left(\tau\right)\right\}d\tau\right\} \leq$$

$$\leq \frac{\varepsilon\left(s\right)}{2}\left(2M\left(s\right) - 1\right).$$

This together with (14.16) gives the required estimate

$$\left|r\left(s\right)\left(t\right) - \overline{r}\left(s\right)\left(t\right)\right| \leq \varepsilon\left(s\right)M\left(s\right),$$

what ends the proof. $\square$

At the moment we have to make a remark that the last result is strongest known one concerning the existence of continuous selections in functional spaces.

4. Retractions of solution sets

This section is devoted to the retraction property of the sets

$$\mathcal{F}\left(s\right) = \left\{u' : u \in \mathcal{R}\left(s\right)\right\}$$

describing the derivatives of all possible solutions to the Cauchy problem (14.3)

$$\begin{cases} u' \in F\left(t, u, s\right), \\ \quad u\left(0\right) = \zeta\left(s\right) \end{cases}.$$

We shall impose slightly stronger hypotheses on $F : I \times X \times S \longrightarrow cl\left(X\right)$ than that of the Theorem 63, namely;

i_r F is $\mathcal{L} \otimes \mathcal{B}\left(X \times S\right) - measurable$ in $\left(t, x, s\right)$;

ii_r for any $\left(t, x\right)$ the mapping $s \longrightarrow F\left(t, x, s\right)$ is continuous;

iii_r there exists an $l \in L^1\left(I, R^+\right)$ such that for every $x_1, x_2 \in X$ and any $s \in S$ the inequality

$$d_H\left(F\left(t, x_1, s\right), F\left(t, x_2, s\right)\right) \leq l\left(t\right)\left|x_1 - x_2\right| \quad a.e. \; in \; I;$$

iv_r there exists $p \in L^1\left(I, R\right)$ such that for every $\left(s, x\right) \in S \times X$

$$\sup\left\{\left|u\right| : u \in F\left(t, x, s\right)\right\} \leq p\left(t\right) \quad a.e. \; in \; I.$$

First of all notice that $\mathcal{F}\left(s\right) = \mathcal{F}ixK\left(s, u\right)$,where

$$K\left(s, u\right) = \left\{w \in L^1\left(I, X\right) : w\left(t\right) \in F\left(t, \mathcal{I}\left(\zeta\left(s\right), u\right)\left(t\right), s\right) \quad a.e. \; in \; I\right\},$$

$$\mathcal{I}\left(\zeta, u\right)\left(t\right) = \zeta + \int_0^t u\left(\tau\right)d\tau.$$

This makes the examination of the multifunction $s \longrightarrow \mathcal{F}\left(s\right)$ to be reliable on the Theorem 57.

THEOREM 64. *Assume that $F : I \times X \times S \longrightarrow cl(X)$ satisfies conditions $i_1 - iv_r$ while $\zeta : S \longrightarrow X$ is continuous. Then each $\mathcal{F}(s)$ is a retract of $L^1(I, X)$. Moreover, it can be found a continuous mapping $r : S \times L^1(I, X) \longrightarrow L^1(I, X)$ which, for every $s \in S$, establishes the retraction of $L^1(I, X)$ onto $\mathcal{F}(s)$.*

PROOF. Introduce the space $L^1(I, X, \mu_0)$ of Bochner integrable functions $u : I \longrightarrow X$ with the equivalent Bielecki norm

$$\|u\|_0 = \int_0^1 |u(t)| \, \mu_0(dt),$$

where μ_0 is the absolutely continuous measure on I defined by

$$d\mu_0 = \exp(-2r(t)), \qquad r(s)(t) = \int_0^t l(s)(\tau)\mu_0(d\tau).$$

We claim that the mapping $K_0 : S \times L^1(I, X, \mu_0) \longrightarrow dcl(I, X, \mu_0)$ given by

$$K_0(s, u) = \{w \in L^1(I, X, \mu_0) : w(t) \in F(t, \mathcal{I}(\zeta(s), u)(t), s) \quad a.e. \ in \ I\}$$

satisfies all hypothesis of the Theorem 57.

The Condition 3 is, by ii_r, obvious.

To see that the Condition 4 holds fix $s \in S$, $\varepsilon > 0$, $u_1, u_2 \in L^1(I, X, \mu_0)$ and $w_1 \in K_0(s, u_1)$. By the Proposition 54 there is $w_2 \in K_0(s, u_2)$ such that

$$|w_1(t) - w_2(t)| < d(w_1(t), F(t, \mathcal{I}(\zeta(s), u_2)(t), s)) + \varepsilon \quad a.e. \ in \ I.$$

Then *a.e. in I* we have

$$|w_1(t) - w_2(t)| < l(t) \left| \int_0^t u_1(\tau) \, d\tau - \int_0^t u_2(\tau) \, d\tau \right| dt + \varepsilon \leq$$

$$\leq l(t) \int_0^t |u_1(\tau) - u_2(\tau)| \, d\tau + \varepsilon$$

and hence

$$\|w_2 - w_1\|_0 = \int_0^1 |w_2(t) - w_1(t)| \, \mu_0(dt) \leq$$

$$\leq \int_0^1 \exp\left(-2r\left(t\right)\right) l\left(t\right) \left(\int_0^t \left|u_2\left(\tau\right) - u_1\left(\tau\right)\right| d\tau\right) dt + \varepsilon =$$

$$= -\frac{1}{2} \int_0^1 \left(\exp\left(-2r\left(t\right)\right)\right)'\left(t\right) \left(\int_0^t \left|u_2\left(\tau\right) - u_1\left(\tau\right)\right| d\tau\right) dt + \varepsilon.$$

Integrating by parts we end up with

$$\left\|w_2 - w_1\right\|_0 \leq -\frac{1}{2}\exp\left(-2r\left(1\right)\right)\int_0^1 \left|u_2\left(t\right) - u_1\left(t\right)\right| dt+$$

$$+\frac{1}{2}\int_0^1 \exp\left(-2r\left(t\right)\right)\left|u_2\left(t\right) - u_1\left(t\right)\right| dt + \varepsilon \leq \frac{1}{2}\left\|u_2 - u_1\right\|_0 + \varepsilon.$$

Since ε is arbitrarily small then

$$d_H\left(w_1, K_0\left(s, u_2\right)\right) \leq \frac{1}{2}\left\|u_2 - u_1\right\|_0.$$

But the role of u_1 and u_2 is symmetric. Therefore

$$d_H\left(K_0\left(s, u_1\right), K_0\left(s, u_2\right)\right) \leq \frac{1}{2}\left\|u_2 - u_1\right\|_0.$$

Now the Theorem 57 applied to the space $L^1\left(I, X, \mu_0\right)$ gives the desired result. □

CHAPTER 15

Decomposable analysis

Let T be a compact metrizable space with a $\sigma - field$ $\mathcal{L}$ of Borel measurable sets given by a probabilistic Lebesgue measure μ, i.e. $\mu(T) = 1$. Let us recall that any closed subset $K \subset L^1(T, X, \mu) = L^1(T, X)$ is decomposable iff there is a measurable multifunction $P : T \longrightarrow cl(X)$ such that

$$K = \{u \in L^1(T, X) : u(t) \in P(t) \ a.e. \ in \ T\}.$$

As we have already shown the decomposable sets are widely used and they are in many situations a substitute of convexity. But up-to now there is no analogue of convex functions. In this chapter we define decomposable mappings and propose a kind of "decomposable analysis" which shall examine such objects.

1. Weak closures of decomposable sets

In a fundation of this section lies a generalization of the works due to Olech [177] and Tranh Cao Nguyen [166] who considered a characterization of weak* closure of decomposable sets $K \subset L^1(T, R^l)$ and $l.s.c.$ of integral functionals $\mathbf{I}(u) = \int_T f(t, u(t)) \mu(dt)$ with $f : T \times R^l \longrightarrow \overline{R} = R \cup \{\infty\}$. Recall that any function $u \in L^1(T, R^l)$ we can identify with a measure

$$m = m_u \in \mathcal{M}_a(T, R^l) \subset \mathcal{M}(T, R^l) \subset \left(L^\infty(T, R^l)\right)^*$$

given by

$$m(A) = \int_A u(t)\mu(dt).$$

For any $z \in L^\infty(T, R^l)$ and $m \in \left(L^\infty(T, R^l)\right)^*$ we have a pairing

(15.1)
$$\langle m, z \rangle = \int_T \langle z(t), m(dt) \rangle.$$

Within the above identification one can write that

$$\langle z, u \rangle = \langle m_u, z \rangle = \int_T \langle z(t), u(t) \rangle \, \mu(dt)$$

for any $z \in L^\infty(T, R^l)$ and $u \in L^1(T, R^l)$.

Let $\mathcal{Z} \subset L^\infty(T, R^l)$ be a linear subspace with it's own topology given by a norm $\|z\|_{\mathcal{Z}}$, not necessarily equivalent to that of $L^\infty(T, R^l)$. About $\mathcal{Z}$ we shall impose the following assumption

CONDITION 5. *a)* $(\mathcal{Z}, \|z\|_{\mathcal{Z}})$ *is a Banach space*
b) *For each* $m \in \mathcal{M}(T, R^l)$

$$\sup\{|\langle m, z \rangle| : \|z\|_{\mathcal{Z}} \le 1\} = \|m\|.$$

As an example of a subspace $\mathcal{Z}$ fulfilling the Condition 5 the reader may think about $L^\infty(T, R^l)$, $C(T, R^l)$, $AC(T, R^l)$ or any $C^k(I)$. Also note that withing the identification of any $u \in L^1(T, R^l)$ with $m_u \in \mathcal{M}(T, R^l)$ we then have

(15.2) $$\sup\{|\langle z, u \rangle| : \|z\|_{\mathcal{Z}} \le 1\} = \|u\|_1.$$

Observe that the Condition 5 means in particular that

$$\left(L^\infty(T, R^l)\right)^* \subset \mathcal{Z}^*.$$

Moreover, $\mathcal{Z}$ is total for $\left(L^\infty(T, R^l)\right)^*$, i.e. if for some $m \in \left(L^\infty(T, R^l)\right)^*$ we have $\langle m, z \rangle = 0$ for each $z \in \mathcal{Z}$ then $m = 0$. In particular $\mathcal{Z}$ is total for $L^1(T, R^l)$ and for $\mathcal{M}(T, R^l)$. Therefore in each of spaces

$$L^1(T, R^l) \subset \mathcal{M}(T, R^l) \subset \left(L^\infty(T, R^l)\right)^*$$

we may consider $\mathcal{Z}$−topology introducing neighbourhoods of $m_0 \in \left(L^\infty(T, R^l)\right)^*$ given by

$$V(m_0, \varepsilon, z_1, ..., z_n) = \left\{ m \in \left(L^\infty(T, R^l)\right)^* : |\langle m - m_0, z_i \rangle| < \varepsilon, \ i = 1, ..., n \right\}.$$

The convergence of a net $\{m_\alpha\} \in \left(L^\infty(T, R^l)\right)^*$ to m in $\mathcal{Z}$−topology, denoted by $m_\alpha \xrightarrow{\mathcal{Z}} m$, means that for every $z \in \mathcal{Z}$ we have $\langle m_\alpha, z \rangle \longrightarrow \langle m, z \rangle$. As we already have said the $\mathcal{Z}$−topology can be restricted to $L^1(T, R^l) \subset \mathcal{M}(T, R^l) \subset \left(L^\infty(T, R^l)\right)^*$. In particular for a net $\{u_\alpha\} \subset L^1(T, R^l)$ we write $u_\alpha \xrightarrow{\mathcal{Z}} m$ whenever $m_\alpha = m_{u_\alpha} \xrightarrow{\mathcal{Z}} m$. In other words $\langle z, u_\alpha \rangle \longrightarrow \langle m, z \rangle$ for every $z \in \mathcal{Z}$. Notice that for $\mathcal{Z} = L^\infty(T, R^l)$ the $\mathcal{Z}$−topology leads simply to the weak convergence, while for $\mathcal{Z} = C = C(T, R^l)$ to the weak* one. So the C−topology in $\mathcal{M}(T, R^l)$ coincides with the weak* one. In both cases the limit of any $\mathcal{Z}$−convergent net $\{u_\alpha\} \subset L^1(T, R^l)$ is a measure. This situation occurs to be, under the Condition 5, general.

PROPOSITION 64. *Assume that $\mathcal{Z} \subset L^\infty(T, R^l)$ satisfies the Condition5. Then $\mathcal{M}(T, R^l)$ is closed in $\mathcal{Z}$—topology. In particular the limit of any $\mathcal{Z}$—convergent net $\{u_\alpha\} \subset L^1(T, R^l)$ is a measure. Moreover, in one dimensional case, if all $u_\alpha \in L^1(T)$ are nonnegative functions and $u_\alpha \xrightarrow{\mathcal{Z}} m$ then $m \geq 0$.*

PROOF. To see the first statement take $\{m_\alpha\} \subset \mathcal{M}(T, R^l)$ such that $m_\alpha \xrightarrow{\mathcal{Z}} m$. Employing the Banach-Steinhaus Boundedness Principle and the Condition 5 one can observe that any $\mathcal{Z}$—convergent net $\{m_\alpha\}$ is bounded. Therefore by the Banach-Alaoglu Theorem it contains a subnet convergent weakly* in $\mathcal{M}(T, R^l)$ and $\left(L^\infty(T, R^l)\right)^*$ to a measure, which therefore has to coincide with m. In the real case the measure $m \geq 0$, since it is a limit of a nonnegative subnet. $\qquad\square$

Provided above discussion allows us to make a consideration concerning the closure $cl_{\mathcal{Z}}K$ of a subset $K \in dec(T, R^l)$ in $\mathcal{Z}$—topology. We shall use for every $z \in \mathcal{Z}$ a denotation

$$\psi_z = ess \sup \left\{ \langle z(\cdot), u(\cdot) \rangle : u \in K \right\}.$$

THEOREM 65. *Let $K \in dec(T, R^l)$. Then the formula*

$$(15.3) \qquad cl_{\mathcal{Z}}K = \bigcap_{z \in \mathcal{Z}} \left\{ m \in \mathcal{M}(T, R^l) : \langle m, z \rangle \leq \int_T \psi_z(t)\mu(dt) \right\}$$

holds.

PROOF. Denote by K_0 the right-hand side of the formula (15.3). If $u_\alpha \xrightarrow{\mathcal{Z}} m$ for a net $\{u_\alpha\} \subset K$ then for every $z \in \mathcal{Z}$ we have $\langle z, u_\alpha \rangle \longrightarrow \langle m, z \rangle$. But $\langle z(t), u_\alpha(t) \rangle \leq \psi_z(t)$ a.e. in T and hence

$$\langle z, u_\alpha \rangle \leq \int_T \psi_z(t)\mu(dt).$$

Therefore for every $z \in \mathcal{Z}$

$$\langle z, m \rangle \leq \int_T \psi_z(t)\mu(dt),$$

what justifies that

$$cl_{\mathcal{Z}}K \subset K_0.$$

For the opposite inclusion take any $m \in K_0$ and let $V(m, \varepsilon, z_1, ..., z_n)$ with $\{z_1, ..., z_n\} \subset \mathcal{Z}$ be an arbitrary but fixed neighbourhood of m. It is sufficient to show that

$$V(m, \varepsilon, z_1, ..., z_n) \cap K \neq \emptyset.$$

For this purpose set
$$D = \{x = (\langle z_1, u \rangle, ..., \langle z_n, u \rangle) : u \in K\} \subset R^n.$$
By the Lapunov Theorem the set D is convex. We shall prove that

(15.4) $$\tilde{z} = (\langle z_1, m \rangle, ..., \langle z_n, m \rangle) \in clD.$$

Indeed,
$$clD = \bigcap_{a \in R^n} \{x \in R^n : \langle x, a \rangle \leq c_D(a)\}.$$
But for every $a = (a_1, ..., a_n) \in R^n$ the support function can be evaluated, by the Lemma 6, as follows
$$c_D(a) = \sup\{\langle a, x \rangle : x \in D\} =$$
$$= \sup\left\{\sum_{i=1}^{n} a_i \int_T \langle z_i(t), u(t) \rangle \mu(dt) : u \in K\right\}$$
$$= \sup\left\{\int_T \left\langle \sum_{i=1}^{n} a_i z_i(t), u(t) \right\rangle \mu(dt) : u \in K\right\} =$$
$$= \int_T \psi_{\sum_{i=1}^{n} a_i z_i}(t) \mu(dt).$$
Hence for each $a \in R^n$ we have
$$\langle a, \tilde{z} \rangle = \sum_{i=1}^{n} a_i \langle z_i, m \rangle = \left\langle \sum_{i=1}^{n} a_i z_i, m \right\rangle \leq \int_T \psi_{\sum_{i=1}^{n} a_i z_i}(t) \mu(dt) = c_D(a),$$

which shows (15.4). Having that it is a simple matter to check that for every $\varepsilon > 0$ there is $u \in K$ with the property that for each $i = 1, ..., n$
$$|\langle z_i, u - m \rangle| < \varepsilon.$$
In other words
$$u \in V(m, \varepsilon, z_1, ..., z_n) \cap K.$$
This completes the proof.   □

In certain situations the formula (15.3) can be obtained by a countable intersection. This occures, for example, if $\mathcal{Z}$ satisfies the following

CONDITION 6. *There exists $\mathcal{Z}_0 \subset L^\infty(T)$ satisfying the Condition 5 such $\mathcal{Z}_0\mathcal{Z} \subset \mathcal{Z}$.*

From the above condition we derive

PROPOSITION 65. *Assume that $\mathcal{Z}$ together with $\mathcal{Z}_0$ satisfy the Conditions 5 and 6. Then*

(1) For given $\{m_\alpha\} \subset \mathcal{M}(T, R^l)$ such that $m_\alpha \xrightarrow{\mathcal{Z}} m_0$ and every $z \in \mathcal{Z}$ the net $\langle z(\cdot), m_\alpha \rangle \xrightarrow{\mathcal{Z}_0} \langle z(\cdot), m_0 \rangle$;

(2) $\mathcal{Z}$-topology and $C\mathcal{Z}$-topology in $\mathcal{M}(T, R^l)$ are equivalent;

(3) For every $z \in \mathcal{Z}$ the set

$$M_z = \left\{ m \in \mathcal{M}(T, R^l) : \langle m, z \rangle \le \int_T \psi_z(t)\mu(dt) \right\}$$

is C-closed.

PROOF. For any $z_0 \in \mathcal{Z}_0$ an easy calculation

$$\int_T z_0(t) \langle z(t), m_\alpha(dt) \rangle = \int_T \langle z_0(t) z(t), m_\alpha(dt) \rangle \longrightarrow$$

$$\int_T \langle z_0(t) z(t), m_0(dt) \rangle = \int_T z_0(t) \langle z(t), m_0(dt) \rangle,$$

shows that $\langle z(\cdot), m_\alpha \rangle \xrightarrow{\mathcal{Z}_0} \langle z(\cdot), m_0 \rangle$. Using the Proposition 64 we conclude that the net $\{\langle z(\cdot), m_\alpha(\cdot) \rangle\}$ contains a weakly* convergent subnet $\{\langle z(\cdot), m_{\widetilde{\alpha}} \rangle\}$ and this implies that for every $c \in C$

$$\langle c(\cdot) z(\cdot), m_{\widetilde{\alpha}} \rangle \xrightarrow{\mathcal{Z}_0} \langle c(\cdot) z(\cdot), m_0 \rangle.$$

Hence

$$\langle z(\cdot), m_{\widetilde{\alpha}} \rangle \xrightarrow{C} \langle z(\cdot), m_0 \rangle$$

and therefore we get that $\mathcal{Z}$-topology and $C\mathcal{Z}$-topology are equivalent and M_z is C-closed.   □

COROLLARY 15. *Assume that $\mathcal{Z}$ satisfies the Conditions 5 and 6. Let $K \in dec(T, R^l)$. Then there exist denumerable sets $\{z_n\} \subset \mathcal{Z}$ and $\{\psi_n\} \subset L^1(T)$ such that*

$$(15.5) \qquad cl_{\mathcal{Z}} K = \bigcap_{n=1}^{\infty} \left\{ m \in \mathcal{M}(T, R^l) : \langle m, z_n \rangle \le \int_T \psi_n(t)\mu(dt) \right\}$$

PROOF. By the Proposition 65 all sets

$$M_z = \left\{ m \in \mathcal{M}(T, R^l) : \langle m, z \rangle \le \int_T \psi_z(t)\mu(dt) \right\}$$

are C-closed. Denote by $B(j)$ the closed ball in $\mathcal{M}(T, R^l) = C^*(T, R^l)$ of radius j. Then the $C-$ topology in $B(j)$ is metrizable and therefore there exist a denumerable set $\Phi_j \subset \mathcal{Z}$ such that

$$B\left(j\right) \cap cl_{\mathcal{Z}} K = B\left(j\right) \cap \bigcap_{z \in \Phi_j} \left\{ m \in \mathcal{M}(T, R^l) : \langle m, z \rangle \leq \int_T \psi_z(t) \mu(dt) \right\}.$$

We may also suppose that $\{\psi_z\}_{z \in \Phi_j} \subset L^1\left(T\right)$. Putting $\{z_n\} = \bigcup_{j=1}^{\infty} \Phi_j$

and $\{\psi_n\} = \bigcup_{j=1}^{\infty} \{\psi_z\}_{z \in \Phi_j}$ we obtain required denumerable sets such that (15.5) holds. $\square$

THEOREM 66. *A set* $K \in dec(T, R^l)$ *is* $\mathcal{Z}-$*closed if and only if* $K = K_P$, *where* $P : T \longrightarrow clco\left(R^l\right)$ *has a representation*

$$P\left(t\right) = \bigcap_{n=1}^{\infty} \left\{ x \in R^l : \langle x, z_n\left(t\right)\rangle \leq \psi_n(t) \ \ a.e. \ in \ T \right\},$$

where $\{z_n\} \subset \mathcal{Z}$ *and* $\{\psi_n\} \subset L^1\left(T\right)$.

PROOF. If $K \in dec(T, R^l)$ is $\mathcal{Z}-$ closed then invoking the formula (15.5) we have a required representation. On the other hand assume that $K = K_P$ has the prescribed form. We shall show that it is $\mathcal{Z}-$ closed. For this purpose take any $u_\alpha \xrightarrow{\ \mathcal{Z}\ } u_0$, $\{u_\alpha\} \subset K$ and fix arbitrarily $n \in N$. Then

$$0 \leq \psi_n\left(\cdot\right) - \langle z_n\left(\cdot\right), u_\alpha\left(\cdot\right)\rangle \xrightarrow{\ \mathcal{Z}_0\ } \psi_n\left(\cdot\right) - \langle z_n\left(\cdot\right), u_0\left(\cdot\right)\rangle$$

and hence for each $n \in N$ the inequality

$$\langle z_n\left(t\right), u_0\left(t\right)\rangle \leq \psi_n\left(t\right)$$

is fulfilled *a.e.* in T. Thus $u_0 \in K$. $\square$

2. Decomposable mappings

DEFINITION 19. *A mapping* $k : L^p(T, X) \longrightarrow L^s(T, R) = L^s(T)$ *we shall call decomposable iff for any* $u, v \in K$ *and* $A \in \Sigma$ *one has*

$$k\left(\chi_A u + (1 - \chi_A)\right) \leq \chi_A k(u) + (1 - \chi_A) k(v).$$

As an example of such mappings we give the Niemytski operator

$$N(u)(\cdot) = f(\cdot, u(\cdot)),$$

where $f : T \times X \longrightarrow R$ is a given $\mathcal{L} \otimes B$-mesurable function satisfying

$$|f(t, u(t))| \leq a(t) + M\, |u(t)|^{\frac{p}{s}},$$

with given $a \in L^s(T)$ and $M \geq 0$.

On the other hand one can easily check that the mapping k_0 from $L^1(I)$ into itself given by $k_0(u) = 0$ for $u \neq 0$ and $k_0(0) = 1$ is decomposable and for $u = \chi_A$, $v = 1 - \chi_A$ with $0 < \mu(A) < \mu(T)$ we have

$$k_0\left(\chi_A u + (1 - \chi_A)\right) < \chi_A k_0(u) + (1 - \chi_A)k_0(v).$$

A purpose to consideration of decomposable mappings comes from modern calculus of variation. In many situations concerning integral functionals

$$\mathbf{I}(u) = \int_T f(t, u(t))\mu(dt)$$

the continuity or lower semicontinuity can be characterized by properties of integrand function. This can be done for strong, weak, weak*-topologies in $L^1(T)$, strong and weak topologies in $L^p(T)$. The necessary and sufficient conditions in all just mentioned situations look similarly and our attempt is to unify in case of certain kinds of weak topologies all these results.

Actually Niemytski operators satisfy condition

$$N\left(\chi_A u + (1 - \chi_A)\right) = \chi_A N(u) + (1 - \chi_A)N(v)$$

and this condition plays similar role as affine mappings in convex analysis.

DEFINITION 20. *A mapping* $k : L^p(T, X) \longrightarrow L^s(T, R)$ *we shall call affine decomposable iff for any* $u, v \in K$ *and* $A \in \Sigma$ *we have*

$$(15.6) \qquad k\left(\chi_A u + (1 - \chi_A)\right) = \chi_A k(u) + (1 - \chi_A)k(v).$$

In this place we have to say that in the decomposable analysis a situation is different from that of the convex analysis. In the convex case there are *l.s.c.* and convex functions $f : X \longrightarrow \overline{R} = R \cup \{\infty\}$ such that for some $x, y \in X$ and $\lambda \in (0, 1)$ is

$$f\left(\lambda x + (1 - \lambda)y\right) < \lambda f(x) + (1 - \lambda)f(y),$$

so in the decomposable case, as we will see below, the assumption of *l.s.c.* in weak topologies forces (15.6). Also at the moment is not obvious whether each affine decomposable mapping is a Niemytski operator. This is an open question and it seems to rely on the topology on $L^p(T, X)$.

PROPOSITION 66. *A mapping* $k : L^p(T, X) \longrightarrow L^s(T, R)$ *is decomposable iff its* $L - epigraph$

$$epi(k) = \{(u, v) \in L^s(T, R) \times L^p(T, X) : v(t) \geq k(u)(t) \text{ a.e. in } T\}$$

is decomposable.

PROOF. Take $(u_1, v_1), (u_2, v_2) \in K$ and arbitrary $A \in \mathcal{L}$. Then $v_i(t) \geq k_i(u)(t)$ $a.e.$ in T, $i = 1, 2$ and therefore

$$k(\chi_A u_1 + (1 - \chi_A)u_2) \leq$$
$$\leq \chi_A k(u_1) + (1 - \chi_A)k(u_2) \leq \chi_A v_1 + (1 - \chi_A)v_2,$$

what shows that

$$\chi_A(u_1, v_1) + (1 - \chi_A)(u_2, v_2) =$$
$$= (\chi_A u_1 + (1 - \chi_A)u_2, \chi_A v_1 + (1 - \chi_A)v_2) \in K.$$

On the other hand for any $u_1, u_2 \in L^p(T)$ and $A \in \mathcal{L}$ we have

$$(\chi_A u_1) + (1 - \chi_A)u_2), \chi_A k(u_1) + (1 - \chi_A)k(u_2)) =$$
$$= \chi_A(u_1), k(u_1)) + (1 - \chi_A)(u_2), k(u_2)) \in K,$$

what shows that

$$k(\chi_A u_1 + (1 - \chi_A)u_2) \leq \chi_A k(u_1) + (1 - \chi_A)k(u_2).$$

$\square$

3. Integral functionals

The considerations from two last sections we are going to apply in examination of the lower semicontinuity in $\mathcal{Z}-$ topology of integral functionals having form

$$\mathbf{I}(u) = \int_T k(u)\,\mu(dt),$$

where $k : L^1(T, R^l) \longrightarrow L^1(T, \overline{R})$ is a decomposable mapping. Such functional $\mathbf{I}$ defines the epigraph

(15.7) $epi(\mathbf{I}) = E = \left\{ (u, r) \in L^1(T, R^l) \times R : r \geq \mathbf{I}(u) \right\},$

while the mapping k is described by $L - epigraph$

(15.8)
$Epi(k) = K = \left\{ (u, v) \in L^1(T, R^l \times R) : v(t) \geq k(u)(t) \quad a.e. \text{ in } T \right\}.$

On $L^1(T, R^l \times R)$ we shall consider $\mathcal{Z} \times C-$topology, where $C = C(T)$. In our case

$$\mathcal{M}(T) = C^*(T)$$

and therefore the weak$^*-$topology on $L^1(T) \subset \mathcal{M}(T)$ coincides with $C-$topology.

We shall assume the following

CONDITION 7. *There is at least one* $u_0 \in L^1(T, R^l)$ *such that* $k(u_0) \in L^1(T, R)$.

For decomposable K given by (15.8) the formula (15.3) reads as follows

$$cl_{Z \times C} K =$$

$$= \bigcap_{(z,c) \in Z \times C} \left\{ (m, m_0) \in \mathcal{M}(T, R^l \times R) : \langle m, z \rangle + \langle m_0, c \rangle \leq \int_T \psi_{(z,c)}(t) \mu(dt) \right\}$$

and in this representation we have to make some reduction concerning possible values of c. Invoking the Condition 7 we get that for every $w \in R^+$ the points

$$(u_0, k(u_0) + w) \in K.$$

Fix $(z, c) \in Z \times C$. Hence for every $w \geq 0$ relation

$$\langle z, u_0 \rangle + \langle k(u_0) + w, c \rangle \leq \int_T \psi_{(z,c)}(t) \mu(dt)$$

holds. Dividing both sides by $w > 0$ and next passing to the limit with $w \longrightarrow \infty$ we get $c(t) \leq 0$ for all $t \in T$. Further reductions we obtain from an observation that the mapping

$$c \longrightarrow \psi_{(z,c)} = ess \sup \{ \langle z, u \rangle + \langle v, c \rangle : (u, v) \in K \}$$

is continuous (if is finite) and we may therefore assume that

(15.9) $$c(t) < 0 \quad for \ all \ t \in T.$$

Let

$$C_+ = \{ c \in C : c(t) > 0 \quad for \ all \ t \in T \}.$$

So we get

$$cl_{Z \times C} K =$$

$$= \bigcap_{z \in Z, c \in C_+} \left\{ (m, m_0) \in \mathcal{M}(T, R^l \times R) : \langle m, z \rangle - \langle m_0, c \rangle \leq \int_T \psi_{(z,c)}(t) \mu(dt) \right\}.$$

Employing the same arguments as for the Corollary 15 and the Theorem 66 we have

THEOREM 67. *Let $K \in dec(T, R^l)$ be given by (15.8). Then there exist denumerable sets $\{z_n\} \subset Z$, $\{c_n\} \in C_+$ and $\{\psi_n\} \subset L^1(T)$ such that*

$$cl_{Z \times C} K =$$

$$= \bigcap_{n=1}^{\infty} \left\{ (m, m_0) \in \mathcal{M}(T, R^l \times R) : \langle m, z_n \rangle - \langle m_0, c_n \rangle \leq \int_T \psi_n(t) \mu(dt) \right\}.$$

Moreover, K is $\mathcal{Z}-$closed if and only if $K = K_P$, where $P : T \longrightarrow$ clco $\left(R^l\right)$ has a representation

$$P(t) = \bigcap_{n=1}^{\infty} \left\{ (x, y) \in R^l \times R : \langle x, z_n(t) \rangle - y c_n(t) \leq \psi_n(t) \quad a.e. \ in \ T \right\},$$

Now we present a result concerning the lower seicontinuity in $\mathcal{Z}-$topology of integral functionals

$$\mathbf{I}(u) = \int_T k(u)\, \mu(dt),$$

where the integrand is given by a decomposable mapping $k : L^1(T, R^l) \longrightarrow L^1(T, \overline{R})$

THEOREM 68. *Assume that $\mathcal{Z}$ satisfies Conditions 5 and 6, while a decomposable mapping $k : L^1(T, R^l) \longrightarrow L^1(T, \overline{R})-$ Condition 7. Then for $\mathbf{I}(u) = \int_T k(u)\, \mu(dt)$ following conditions are equivalent:*

(1) *$\mathbf{I} : L^1(T, R^l) \longrightarrow R \cup \{+\infty\}$ is l.s.c. in $\mathcal{Z}-$topology;*
(2) *the L-epigraph Epi $(k) = K$ given by (15.8) is $\mathcal{Z} \times C-$closed;*
(3) *there are functions $z_n \in \mathcal{Z}$, $c_n \in C_+$ and $\psi_n \in L^1(T)$, $n = 1, 2, ...,$ such that for each $u \in L^1(T, R^l)$ the relation*

$$k(u)(t) = \sup\left(-\psi_n(t) + c_n(t) \langle z_n(t), u(t) \rangle \right)$$

holds a.e. in T;
(4) *the epigraph epi $(\mathbf{I}) = E$ given by (15.7) is $\mathcal{Z} \times R-$closed and convex.*

PROOF. (1) $\Longrightarrow$ (2)   Let $\{(u_\alpha, v_\alpha)\} \subset K$ be a net which is $\mathcal{Z} \times C-$convergent to (u_0, v_0). We need to show that

(15.10) $v_0(t) \geq k(u_0)(t) \quad a.e. \ in \ T.$

But for every α we have

$$v_\alpha(t) \geq k(u_\alpha)(t) \quad a.e. \ in \ T$$

and hence

$$\limsup \|v_\alpha - k(u_\alpha)\|_1 = \limsup \int_T (v_\alpha(t) - k(u_\alpha))\, \mu(dt) \leq$$

$$\leq \int_T v_0(t)\, \mu(dt) - \mathbf{I}(u_0).$$

This implies that the net $\{v_\alpha - k(u_\alpha)\}$ is bounded in $L^1(T)$ and therefore C−compact. So passing to a subnet, we may assume that $k(u_\alpha)$ is C−convergent to a measure m. Thus

$$v_\alpha - k(u_\alpha) \xrightarrow{C} v_0 - m.$$

But $v_\alpha - k(u_\alpha) \geq 0$, so $v_0 - m \geq 0$. Therefore for the absolute part m_a of m we have

(15.11)
$$v_0(t) \geq \frac{dm_a(t)}{dt} \quad a.e. \text{ in } T,$$

while for the singular

$$m_s \leq 0.$$

We claim that

$$\frac{dm_a(t)}{dt} \geq k(u_0)(t) \quad a.e. \text{ in } T,$$

what together with (15.11) shows (15.10). To prove our claim recall that, by the Theorem 67, there are sequences $\{z_n\} \subset \mathcal{Z}$, $\{c_n\} \subset C_+$ and $\{\psi_n\} \subset L^1(T)$ such that

$$cl_{\mathcal{Z} \times C} K =$$

$$= \bigcap_{n=1}^{\infty} \left\{ (m, m_0) \in \mathcal{M}(T, R^l \times R) : \langle m, z_n \rangle - \langle m_0, c_n \rangle \leq \int_T \psi_n(t)\mu(dt) \right\}.$$

Denote

$$P(t) = \bigcap_{n=1}^{\infty} \{(x, r) \in R^l \times R : \langle x, z_n(t) \rangle - rc_n(t) \leq \psi_n(t) \quad a.e. \text{ in } T\}.$$

Fix n and observe that for every α the inequalities

$$\langle u_\alpha, z_n(t) \rangle - v_\alpha c_n(t) \leq \langle u_\alpha, z_n(t) \rangle - k(u_\alpha)(t) c_n(t) \leq \psi_n(t)$$

hold $a.e.$ in T. But, by the Condition 5, $\langle u_\alpha, z_n(t) \rangle \xrightarrow{z_0} \langle u_0, z_n(t) \rangle$ and therefore we may assume that

$$\langle u_\alpha, z_n(t) \rangle \xrightarrow{C} \langle u_0, z_n(t) \rangle.$$

So passing to the limit we obtain that

$$\langle u_\alpha(\cdot), z_n(\cdot) \rangle - v_\alpha(\cdot) c_n(\cdot) \xrightarrow{C} \langle u_0(\cdot), z_n(\cdot) \rangle - \langle m, c_n(\cdot) \rangle$$

and thus

$$\langle u_0(\cdot), z_n(\cdot) \rangle - v_0(\cdot) c_n(\cdot) \leq \langle u_0(\cdot), z_n(\cdot) \rangle - \langle m, c_n(\cdot) \rangle \leq \psi_n(\cdot)$$

From the latter inequality we see that for each n

$$\langle m_s, c_n \rangle = 0.$$

Hence

$$v_\alpha c_n \xrightarrow{C} \frac{dm_a}{d\mu} c_n,$$

what implies that

$$v_0 = \frac{dm_a}{d\mu}$$

Moreover for a.e. $t \in T$ the inequality

$$\langle u_0(t), z_n(t) \rangle - v_0(t) c_n(t) \leq \psi_n(t),$$

is fulfilled. This explains that

$$(u_0(t), v_0(t)) \in P(t) \quad a.e. \; in \; T.$$

Provided arguments justify, in view of the Theorem 67, an equality $K = K_P$. But this gives (15.10) and completes the proof of (2).

$(2) \Longrightarrow (3)$. Assuming that K is $\mathcal{Z} \times C-$closed we can invoke the Theorem 67 to represent K as $K = K_P$, where

$$P(t) = \bigcap_{n=1}^{\infty} \left\{ (x, r) \in R^l \times R : \langle x, z_n(t) \rangle - r c_n(t) \leq \psi_n(t) \quad a.e. \; in \; T \right\}$$

with

$$\{z_n\} \subset \mathcal{Z}, \quad \{c_n\} \subset C_+ \quad and \quad \{\psi_n\} \subset L^1(T).$$

Since for every $u \in L^1(T, R^l)$ the point $(u, k(u)) \in K$ then for any $n \in N$

$$\langle u(t), z_n(t) \rangle - k(u)(t) c_n(t) \leq \psi_n(t) \quad a.e. \; in \; T.$$

Denoting $\widetilde{c}_n = \frac{1}{c_n} > 0$ and $\widetilde{\psi}_n = \frac{1}{c_n} \psi_n$ we obtain

(15.12) $$k(u)(t) \geq -\widetilde{\psi}_n(t) + \widetilde{c}_n(t) \langle u(t), z_n(t) \rangle \quad a.e. \; in \; T.$$

Put

$$\widetilde{k}(u)(t) = \sup \left(-\widetilde{\psi}_n(t) + \widetilde{c}_n(t) \langle u(t), z_n(t) \rangle \right).$$

and notice that (15.12) leads to

$$k(u)(t) \geq \widetilde{k}(u)(t) \quad a.e. \; in \; T.$$

We shall show that

$$k(u)(t) = \widetilde{k}(u)(t) \quad a.e. \; in \; T.$$

For this purpose it sufficies to check that

$$\left(u, \widetilde{k}(u) \right) \in K = K_P.$$

Assuming to a contrary that

$$\left(u, \widetilde{k}(u) \right) \notin K_P$$

we conclude that for some n would be

$$\langle u(t), z_n(t) \rangle - c_n(t)\, \widetilde{k}(u)(t) > \psi_n(t)$$

for t coming from certain set T_0 of positive measure. Hence

$$\widetilde{k}(u)(t) < -\widetilde{\psi}_n(t) + \widetilde{c}_n(t) \langle u(t), z_n(t) \rangle,$$

what contradicts with the definition of $\widetilde{k}(u)$. Hence we have derived 3.

(3) $\Longrightarrow$ (4). Suppose that

$$k(u)(t) = \sup\left(-\psi_n(t) + c_n(t) \langle z_n(t), u(t) \rangle\right),$$

where

$$\{z_n\} \subset \mathcal{Z}, \quad \{c_n\} \subset C_+(T) \quad and \quad \{\psi_n\} \subset L^1(T).$$

We shall show that

$$E = epi\,(\mathbf{I}) =$$

$$= \bigcap_{n=1}^{\infty} \left\{ (u,r) \in L^1(T, R^l) \times R : -r + c_n(t) \langle z_n(t), u(t) \rangle \leq \psi_n(t) \quad a.e.\ in\ T \right\}.$$

Denote the right-hand of the above formula by E_0. If $(u,r) \in E_0$ then for each $n \in N$ we have

$$-\psi_n(t) + c_n(t) \langle z_n(t), u(t) \rangle \leq r \quad a.e.\ in\ T.$$

Thus $k(u)(t) \leq r$, what implies that $(u,r) \in E = epi\mathbf{I}$. Therefore

$$E_0 \subset E.$$

To see the opposite inclusion notice that from the Proposition 4 we get

$$clco_{\mathcal{Z} \times R}(E) =$$

$$= \bigcap_{(z,\zeta) \in \mathcal{Z} \times R} \left\{ (u,r) \in L^1(T, R^l) \times R : \zeta r + \langle z, u \rangle \leq c_E(z, \zeta) \right\} =$$

$$= \bigcap_{(z,c,\zeta) \in \mathcal{Z} \times C \times R} \left\{ (u,r) \in L^1(T, R^l) \times R : \zeta r + \langle cz, u \rangle \leq c_E(cz, \zeta) \right\}.$$

The latter equality follows from the Proposition 65, since $\mathcal{Z}$−topology and $C\mathcal{Z}$−topology are equivalent. Making a similar reduction as in the proof of the Theorem 67 we may assume that the intersection can be taken over $(z, c, \zeta) \in \mathcal{Z} \times C_+ \times R_-$. Therefore, by the positive homogenuity of the support function, we get

$$clco_{\mathcal{Z} \times R}(E) =$$

$$= \bigcap_{(z,c) \in \mathcal{Z} \times C_+} \left\{ (u,r) \in L^1(T, R^l) \times R : -r + \langle cz, u \rangle \leq c_E(cz, -1) \right\} \subset$$

$$\subset \bigcap_{n=1}^{\infty} \left\{ (u,r) \in L^1(T,R^l) \times R : -r + \langle c_n z_n, u \rangle \leq c_E(c_n z_n, -1) \right\}$$

But

$$c_E(c_n z_n, -1) = \sup \left\{ -r + \langle c_n z_n, u \rangle : (u,r) \in E \right\} =$$

$$= \sup \left\{ \int_T (-k(u)(t) + \langle c_n(t) z_n(t), u(t) \rangle) \, \mu(dt) : u \in L^1(T,R^l) \right\} \leq$$

$$\leq \int_T \psi_n(t) \mu(dt).$$

Thus

$$clco_{\mathcal{Z} \times R}(E) \subset$$

$$\subset \bigcap_{n=1}^{\infty} \left\{ (u,r) \in L^1(T,R^l) \times R : -r + c_n(t) \langle z_n(t), u(t) \rangle \leq \psi_n(t) \ a.e. \ in \ T \right\}.$$

So

$$clco_{\mathcal{Z} \times R}(E) = E_0,$$

what gives (4).

(4) $\Longrightarrow$ (1).

It is straightforward. $\square$

REMARK 15. *Actually in the Theorem 68 we have shown that the decomposability assumption for $k : L^1(T,R^l) \longrightarrow L^1(T,\overline{R})$ implies affine decomposability. Moreover, each l.s.c. integral functional $\mathbf{I}$ is a supremum of continuous integral functionals.*

Bibliography

[1] Anichini G., Conti G., Zecca P., Un teorema di selezione per applicazioni multivoche a valori non convessi, Boll. Un. Mat. Ital. C (5), vol. 1 (1986), 315-320.

[2] Andres A. and Górniewicz L., Topological Fixed Point Principles for Boundary Value Problems, Kluwer Acad. Publ., Top. Fixed Point Theory and its Appl., Vol. 1., 2003.

[3] Antosiewicz, H. and Cellina, A., Continuous selections and differential relations, J. Diff. Eq. 19 (1975), 386-398.

[4] Anroszajn N., Le correspondant topologique de l'unicite dans la theorie des equations differentielles, Ann. Math. 43 (1942), 730-738.

[5] Artstein Z., Yet another proof of Lyapunov conexity theorem, Proc. Amer Math. Soc. (1985).

[6] Artstein Z., Parametrized integration of multifuctions with applications to control and optimizations, SIAM J. Control and Opt. 27 (1989).

[7] Artstein Z., Prikry K., Carathéodory Selections and the Scorza-Dragoni Property, JMAA 127 (1987), 540-547.

[8] Aubin J. P., Cellina A., Differential inclusions, Springer Verlag, Berlin, 1984.

[9] Aubin J. P., Frankowska H., Set-Valued Analysis, Birkhaŭser, Boston, Basel, Berlin, 1990.

[10] Aubin J. P., Frankowska H., On inverse function theorems for set-valued maps., J. Math. Pures Appl. 65 (1987), no. 1, 71-89.

[11] Aumann, R.J., Integrals of set-valued functions, J. Math. Anal. Appl., 12 (1965), 1-12.

[12] Bader R., Kryszewski W., Fixed-point index for compositions of set-valued maps with proximally ∞-connected values on arbitrary ANR's, Set-Valued Anal. 2 (1994), 495-480.

[13] Bartuzel G., Fryszkowski A., On existence of solutions of $\nabla u \in F(x, u)$, Proc. of the Fourth Conf. On Numerical Treatment of Ordinary Diff. Eqs., R. Maerz Editor, Berlin 1984;

[14] Bartuzel G., Fryszkowski A., A topological property of the solution set to the Sturm-Liouville differential inclusions, Demonstratio Math. 28 No 4 (1995), 903-914;

[15] Bartuzel G., Fryszkowski A., Abstract differential inclusions with some applications to partial differential ones, Annales Math. Polon. 53 (1991), 67-78;

[16] Bartuzel G., Fryszkowski A., Relaxation of the differential inclusions of the Sturm-Liouville type, Demonstratio Math. 30 No 4 (1997), 953-960.

[17] Bartuzel G., Fryszkowski A., Stability of the principal eigenvalue of the Schroedinger type problems for differential inclusions, Proc. of the International Conference in Nonlinear Analysis, Toruń 1999; Topological Methods in Nonlinear Analysis, vol. 16 (2000), 181-194.

[18] Bartuzel G., Fryszkowski A., A class of retracts in with some applications to differential inclusions, Discussiones Math. 22 (2001), 213-224.

[19] Beer G., Dense selections, J. Math. Anal., Appl. 95(2) (1983), 416-427.

[20] Beer G., Approximate selections for upper semicontinuous convex valued multifunctions, J. Approx. theory, 39(2) (1983), 172-184.

[21] Beer G., On a theorem of cellina for set valued functions, Rocky Mountain J. Math. 18 (1988), 37-47.

[22] Bielwaski R., Górniewicz L., Plaskacz S., Topological Approach to differential inclusions and closed subsets of R^n, Dynam. Report Expositions Dynam. System (new Series) (1992), 225-250.

[23] Blackwell D., The range of certain vector integrals, Proc. Amer. Math. Soc. 2 (1951), 390-395.

[24] De Blasi F. S., Characterisations of certain classes of semicontinuous multifunctions by continuous approximations, J. Math. Anal. Appl. 196(1) (1985), 1-18.

[25] De Blasi F. S., Myjak J., Sur l'existence de selections continues, C. R. Accad. Sci. Paris Sér. I Math. 296(17) (1983), 737-739.

[26] De Blasi F. S., Myjak J., On the solutions sets for differential inclusions, Bull. Pol. Acad. Sci. Math. 12 (1985), 17-23.

[27] De Blasi F. S., Piangiani G., Topological properties of nonconvex differential inclusions, Nonlinear Anal. 20 (1993), 871-894.

[28] De Blasi F. S., Piangiani G., Solution sets of boundary value problems for nonconvex differential inclusions, Topol. Methods Nonlinear Anal. 1 (1993), 303-314.

[29] De Blasi F. S., Myjak J., Continuous selections for weakly Hausdorff lower semicontinuous multifunctions, Proc. Amer. Math. Soc. 93 (1985), 369-372.

[30] Bogatyrev W. A., Fixed Points and properties of solutions of differential inclusions, Izv. Akad. Nauk. SSSR 47 (1983), 895-909. (in Russian)

[31] Bonanno G., Two theorems on the Scorza Dragoni property for multifunctions, Att. Accad. Naz. Lincei Cl. Sci. Fis. Mat. Natur. Rend. 93 (1989), 51-56.

[32] Border K.C., Fixed Point Theorem with applications to economics and Game Theory, Cambridge University Press, London, New York, Melbourne, Sydney, 1985

[33] Borsuk K.,Theory of Retracts, PWN, Warszawa, 1966.

[34] Bressan A., On differential relations with lower semicontinuous right-hand side, J. Diff. Eq. 37 (1980), 89-97.

[35] Bressan A., Colombo G., Extensions and selections of maps with decomposable values, Studia Math. 90 (1988), 69-86.

[36] Bressan A., Cellina A., Fryszkowski A., A class of absolute retracts in spaces of integrable functions, Proceedings of American Math. Soc. 114 (1991), 413-418.

[37] Bressan A., Directionally continuous selections and differential inclusions, Funkcial. Ekvac. 31 (1988), 459-470.

[38] Bressan A., Upper and lower semicontinuous differential inclusions. A unified approach, Controllalitity and Optimal Control (1990), M. Drekker, 21-31.

[39] Bressan A., Solutions of lower semicontinuous differential inclusions on closed sets, Rend. Sem. Mat. Univ. Padova 69 (1983).

[40] Bressan A., Colombo G., Generalized Baire category and differential inclusions in Banach spaces, J. Differential Rquations 76(1) (1988), 135-158.

[41] Brown A. L., Set-valued mappings, continuous selections and approximative selections and metric projections, J. Approx. Theory 57 (1989), 48-68.

[42] Castaing C., Sur les equations differentielles multivoques, C. R. Acad. Sci Apris Sér. I Math. 263 (1966), 63-66

[43] Castaing C., Valadier M., Convex analysis and measurable multifunctions, Lecture Notes in Math. 580, Springer Verlag, Berlin, 1977.

[44] Cellina, A., A fixed point theorem for subsets of L^1, preprint SISSA (1983).

[45] Cellina, A., On the set of solutions to Lipschitzian differential inclusions, Diff. and Int. Equations 1 (1988), 495-500.

[46] Cellina A., Ornelas A., Convexity and closure of the solution set to differential inclusions, Preprint SISSA, 136 M, 1988.

[47] Cellina A., Ornelas A., Representations of the attainable set for Lipschitzian differential inclusions, Rocky Mountains J. of Math. (1988).

[48] Cellina A., Fryszkowski A., Rzeżuchowski T., Upper semicontinuity of Nemytskij Operators, Annali di Matem. pura ed appl. 160 (1991), 321-330.

[49] Cellina A., A selections theorems, Rend. Sem. Mat. Univ. Padova 55 (1976), 99-107.

[50] Cellina A., Colombo R. M., An existence result for differential inclusions with non-converx right-hand side, Funkcial. Ekvac. 32 (1989), 407-416.

[51] Cellina A., Colombo G., Fonda A., Approximate selections and fixed points for upper semicontinuous maps with decomposable values, Proc. Amer. Math. Soc. 98(4) (1986), 663-666.

[52] Cellina A., Marchi M. V., Non-convex perturbations of maximal monote differential inclusions, Israel J. Math. 46(1-2) (1983), 1-11.

[53] Colombo R.M., Fryszkowski A., Rzeżuchowski T., Staicu V., Continuous selection of solution sets of Lipschitzean differential inclusions, Funkcialaj Ekvacioj 34 (1991), 321-330.

[54] Danzer L., Grunbaum B., Klee V., Helly,s theorem and its relatives, Proc. of Symp. in pure Math. VII, AMS, 1963.

[55] Diestel J., Uhl J.J., Vector measures, Mathematical Surveys 15, American Mathematical Society, Providence, 1977.

[56] Deimling K., Multivalued Differential Equations, W. De Gruyter, BErlin, 1992.

[57] Deimling K., On solutions sets of multivalued differential equations, Appl. Anal. 30 (1988), 129-135.

[58] Dentcheva D., Differentiable Selections and Castaing Representation of Multifunctions, JMAA 223 (1988), 371-396.

[59] Dentcheva D., Continuity of multifunctions characterized by Steiner Selections, Nonlinear An. 47 (2001), 1985-1996.

[60] Deutsch F., A survey of metric selections, Contemporary Math. 18 (1983), 49-71

[61] Deutsch F., Kenderov P., Continuous selections and approximative selectionsfor set-valued mappings and applications to metric projections, SIAM J. Math. Anal. 14 (1983), 185-194.

[62] Dunford N, Schwartz J.T, Linear operators, Wiley Classics Library.

[63] Dugundij J., Granas A., Fixed Point Theory I, Monografie Mat, vol. 61, PWN, Warszawa, 1982.

[64] Dzedzej Z., Fixed piont index theory for a class of nonacyclic multivalued maps, Dissertationes Mathematicae, vol. 253, PWN, Warszawa, 1985.

[65] Etienne J., Sur une demonstration du "bang-bang" principle, Bull. Soc. Roy. Sci. Liege 11/12 (1968), 551-556.

[66] Fan K., Fixed point and minimax theorems in locally convex topological linear spaces, Proc. Nat. Acad. Sci. U.S.A. 38 (1952), 271-275.

[67] Fan K., Some properties of convex sets related to fixed point theorems, Math. Ann. 266(4) (1984), 519-537.

[68] Filippov A.F., Classical solutions of differential equations with multivalued right hand side, Yestnik Moscov. Univ. ,Ser. Mat. Mech. Astr., 22,1967,16-26, [English translation: SIAM J. Control. 5, 1967, 609-621].

[69] Filippov V.V., Differetialnye uravnenya s razrywnoy pravoy chastyu, M. Nauka, vol. 224 (1985).

[70] Fonda, A., Some existence results for non-convex valued differential inclusions, M. Ph. Thesis, SISSA, 1986.

[71] Frankowska H., Olech C., R-convexity of the integral of set-valued functions; Contributions to analysis and geometry, Johns Hopkins University Press, Baltimore, 1981, 117-129.

[72] Fryszkowski A., Continuous selections for a class of non-convex multivalued maps, Studia Math. 75 (1983), 163-174.

[73] Fryszkowski, A., Carathéodory type selectors of set-valued maps of two variables, Bull. of Polish Acad. Sci. Vol.XXV, No 1, (1975), 41-46.

[74] Fryszkowski A., Properties of the set of solutions of orientor equation, Proceedings of the 2nd Conference in Diff. Eqs. (1981), 760-763.

[75] Fryszkowski A., The generalization of Cellina's Fixed Point Theorem, Studia Math. 78 (1984), 213-215.

[76] Fryszkowski A., Existence of solutions of functional-differential inclusions in nonconvex case, Annales Math. Polon., 45 (1985).

[77] Fryszkowski A., Carathéodory type selections for some nonconvex multivalued maps and their applications to random multivalued differential-functional eqs., Colloquium. Math. 1983.

[78] Fryszkowski A., Continuous selections of Aumann integrals, Journal of Math. Analysis and Appl. 145 (1990), 431-446.

[79] Fryszkowski A., Carathéodory type selections for Aumann integrals, Modern Optimal Control, Lectures Notes, vol 119 (1990), ed. Emilio O. Roxin, 105-113.

[80] Fryszkowski A., Górniewicz L., Mixed Semicontinuous mappings and their applications to diff. inclusions, Set-valued Analysis vol. 8 (2000), str. 203-217.

[81] Fryszkowski A., Rzeżuchowski T., Continuous version of Fillipov-Ważewski Relaxation Theorem, Journal of Differential Eqs. 94 (1992), 254-265;

[82] Gabor G., Fixed points of set-valued maps with closed proximally ∞-connected values, Discuss. Math. 15 (1995), 163-185.

[83] Gelman B. D., Teorema o nepodviznyh tochkah typa Kakutani dla mnogoznachnyh otobrazeniy, Global. Anal. I Nelinei N. Uravneniva. - Voronezh (1988), 90-105

[84] Górniewicz L., Topological approach to differential inclusions, Topological Methods in Differential Equations And Inclusions, NATO ASI Series C 472 (Granas A., Frigon M., eds), Kluwer Academic Publ., 1995.

[85] Górniewicz L., On the solutions set of differential inclusions, J. Math. Anal. Appl. 113 (1986), 235-244.

[86] Górniewicz L., Remarks on the Lefschetz fixed point theorem, Bull. Polish Acad. Sci. Math. 11 (1973), 993-999.

[87] Górniewicz L., Some consequences of the Lefschetz fixed point theorem for multi-valued maps., Bull. Polish Acad. Sci. Math. 2 (1973), 165-170.

[88] Górniewicz L., Fixed point theorem for multi-valued mappings of approximate ANR's, Fund. Math. 18 (1970), 431-436.

[89] Górniewicz L., Topological Fixed Point Theory of Multivalued Mappings, Kluwer Acad. Publ., 1999.

[90] Górniewicz L., On non-acyclic multi-valued maps of subsets of Euclidean spaces, Bull. Polish Acad. Sci. Math. 5 (1972), 379-385..

[91] Górniewicz L., Topological approach to differential inclusions, Topological Methods in Differential Equations And Inclusions, NATO ASI Series C 472 (Granas A., Frigon M., eds), Kluwer Academic Publ., 1995.

[92] Górniewicz L., Granas A., Fixed point theorems for multi-valued maps of special ANR's, J. Math. Pures Appl. 49 (1970), 381-395.

[93] Górniewicz L., Granas A., Kryszewski W., On the homotopy method in the fixed point index theory for multivalued mappings of compact ANR's, J. Math. Anal. Appl. 191 (1991), 457-473.

[94] Górniewicz L., Marano S. A., On the fixed point set of multi-valued contractions, Rend. Circ. Mat. Palermo 40 (1996), 139-145.

[95] Górniewicz L., Marano S. A., Ślosarski M., Fixed point of contractive multi-valued maps, Proc. Amer. Math. Soc. 124 (1996), 2675-2683.

[96] Górniewicz L., Plaskacz S., periodic solutions of differential inclusions of R^n, Boll. Un. Mat. Ital. A 7 (1993), 409-420.

[97] Górniewicz L., Rozpłoch-Nowakowska G., On the Schauder fixed point theorem, Topology in Nonlinear Analysis Banach Center Publ., vol. 35, Inst. of Math. Polish. Acad. of Sci., Warszawa, 1996.

[98] Granas A., Some theorems in fixed point theory. The Leray-Schauder Index of Lefschetz number, Bull. Polish Acad. Sci. Math. 16 (1968), 131-137.

[99] Granas A., Jaworski J., Some theorems of multivalued maps of subsets of Euclidean space, Bull. Polish Acad. Sci. Math. 6 (1965), 277-283.

[100] Gutev V., Selection theorems under an assumption weaker than lower semi-continuity, Topology Appl. 50 (1993), 129-138.

[101] Haddad G., Monotone trajectories of differential inclusions and functional differential inclusions with memory, Israel J. Math. 39 (1981), no. 1-2, 83-100.

[102] Haddad G., Monotone viable trajectories for functional differential inclusions, J. Differential Equations 42 (1981), no. 1, 1-24.

[103] Haddad G., Topological properties of the sets of solutions for functional differential equations, Nonlinear Anal. 5 (1981),1349-1366.

[104] Haddad G., Lasry J. M., Periodic solutions for functional differential inclusions and fixed points of σ-selectionable correspondences, J. Math. Anal. Appl. 96 (1983), no. 2, 295-312.

[105] Halkin H., Some further generalization of a theorem of Lyapounov, Arch. Rational Mech. Anal. 17 (1964), 272-277.

[106] Halmos P. R., The range of a vector measure, Bull. Amer. Math. Soc. 54 (1948), 416-421.

[107] Hermes, H., Functional analysis and time optimal control, Mathematics in science and engineering 56 (1969).

[108] Hiai F., Umegaki H., Integrals, conditions expectations and martingales of multivalued functions, J. Multiv. Anal. 7 (1977), 149-182.

[109] Himmelberg C.J., van Vleck F.S., Lipschitzian generalized differential equations, Rend. Sem. Mat. Padova, 48 (1972), 159-169.

[110] Himmelberg C.J., Measurable relations. Fund. Math. 87 (1975), 53-72.

[111] Himmelberg C.J., Fixed points of compact multi-valued maps, J. Math. Anal. Appl. 38 (1972), 205-209.

[112] Himmelberg C.J., Parthasarathy T., van Vleck F.S., On measurable relations, Fund. Math. 111 (1981), no. 2, 161-167.

[113] Himmelberg C.J., van Vleck F.S., A note on the solutions sets for differential inclusions, Rocky Mountain J. Math. 12 (1982), 621-625.

[114] Himmelberg C.J., van Vleck F.S., An extension of Brunovsky's Scorza-Dragoni type theorem for unbounded set-valued functions, Math. Slovaca 26 (1976), 47-52.

[115] Himmelberg C.J., van Vleck F.S., Existence of solutions for generalized differential equations with unbounded right-hand side, J. Differential Equations 61 (1986), no. 3, 295-320.

[116] Hu Sh., Papageorgiou N. S., Handbook of Multivalued Analysis, vol. I, Kluwer, 1997.

[117] Hukuhara M., Sur l'application semi-continue dont la valeur est un compact convex, Funkcial Ekvac. 10 (1967), 43-66.

[118] Jacobs M. I., Measurable multivalued mappings and Lusin's Theorem, Trans. Amer. Math. Soc. 134 (1968), 471-481.

[119] Jarnik J., Kurzweil J., On the conditions on right hand sides of differential relations, Casopis Pest. Math. 102 (1977), 334-349.

[120] Jarnik J., Kurzweil J., Extension of a Scorza-Dragoni Theorem to differential relations and functional-differential relations, Comment. Math., special issue in honour of Wladyslaw Orlicz, PWN, Warszawa, 1978, pp. 147-158

[121] Ioffe A. D., On lower semicontinuity of integral functionals I, SIAM J. Control and Opt. 15 (1977), 521-538.

[122] Ioffe A. D., On lower semicontinuity of integral functionals II, SIAM J. Control and Opt. 18, 1978.

[123] Ioffe A. D., Single-valued representations of set-valued mappings II; Applications to differential inclusions, SIAM J. Control. Optim. 21 (1983), 641-651.

[124] Kaczyński H., Olech C., Existence of solutions of orientor fields with nonconvex right-hand side, Annal. Polon. Math. 29 (1974), 61-66.

[125] Kaczyński T., Unbounded multivalued Nemytskii operators in Sobolev spaces and their applications to discontinuous nonlinearity, Rocky Mountain J. Math. 22 (1992), 635-643.

[126] Kakutani S., A generalization of Brouwer's fixed point theorem, Duke Math. J. 8 (1941), 457-459.

[127] Kandilakis D. A., Papageorgiou N. S., on the properties of the Aumann integral with applications to differential inclusions and control systems, Czechoslovak Math. J. 39 (1989), no. 1, 1-15.

[128] Karafiat A., On the continuity of a mapping inverse to a vector measure, Commentationes Mathematicae, 18 (1974/76), 37-43.

[129] Kelley W. G., Periodic solutions of generalized differential equations, SIAM J. Appl. Math. 30 (1976), 70-74.

[130] Kim T., Prikry K., Yannelis N. C., Carathéodory type selections and random fixed point theorems, JMAA 122 (1987), 393-407.

[131] Kingman J. F. C., Robertson A. P., On a theorem of Lyapunov, J. London Math. Soc. 43 (1968), 347-351.

[132] Kisielewicz M., Differential inclusions and optimal control, Polish Sc. Publishers and Kluwer Academic Publishers, Dordrecht, Boston, London, 1991.

[133] Kluvánek I., Knowles G., Vector Measures and Control Systems, North-Holland Publishing Company, Amsterdam, Oxford, American Elsvier Publishing Company, Inc, New York, 1976.

[134] Kucia A., Extending Carathéodory functions, Bull. Polish Acad. Sci. Math. 36 (1988), 593-601.

[135] Kucia A., On the existence of Carathéodory selectors, Bull. Polish Acad. Sci. Math. 32 (1984), 233-241.

[136] Kucia A., Scorza Dragoni type theorems, Fund. Math. 138 (1991), 197-203.

[137] Kucia A., Nowak A., Carathéodory type selectors in Hilbert spaces, Ann. Math. Sil. 14 (1986), 47-52.

[138] Kucia A., Nowak A., Approximate Carathéodory selections, Bull. Polish Acad. Sc. 48 (2000), 81-87.

[139] Kuratowski K., Ryll-Nardzewski C., A general theorem on selectors, Bull. Polish Acad. Sc. 13, 1965, 397-403.

[140] Lapunov A. A., Sur le fonction-vecteurs completement additives, Izv. Akad. Nauk. SSSR Sér Math. 4 (1940), 465-478.

[141] LaSalle, J.P., The time optimal control problem. Contr. to the theory of non linear oscilation, Princeton Univ. Press., Princeton, 1960.

[142] Lasota A., Olech C., On the closedness of the set of trajectories of a control system, Bull. Acad. Polon. Sci. Sér. Math. 14 (1966), 615-621.

[143] Lasota A., Opial Z., An application of the Kakutani-Ky Fan theorem in the theory of ordinary differential equations, Bull. Polish Acad. Sci. Math. 13 (1965), 781-786.

[144] Lasota A., Opial Z., Fixed point theorem fr multivalued mappings and optimal control problems, Bull. Polish Acad. Sci. Math. 16 (1968), 645-649.

[145] Lasota A., Myjak J., Fractals, semifractals and Markov operators, International J. of Bifurcation and Chaos, 9 (1999), 307-325.

[146] Lasry J. M., Robert R., Analyse non lineaire multivoque, Publication no 7611, Centre de Recherche de Mathematique de la Decision Ceremade, Univ. de Paris, Dauphine.

[147] Lasry J. M., Robert R., Degré pour les fonctions multivoques et applications, C. R. Accad. Sci. Paris Sér. I Math. 280 (1975), 1435-1438.

[148] Levinson N., Minimax, Liapunov and "Bang-Bang", J. Differential Equations 2 (1966), 218-241.

[149] Lin W., Continuous selections for set valued mappings, J. Math. Anal. Appl. 188 (1994), 1067-1072.

[150] Lindenstrauss, J., A short proof of Lyapounov's convexity theorem, J. Math. and Mech. 15 (1966), 971-972.

[151] Lojasiewicz St., The existence of solutions for lower semicontinuous orientor fieid, Buli. Acad. Poi. Sc. 28, 1980, 483-487.

[152] Macki J. W., Nistri P., Zecca P., An existence of periodic solutions to nonautonomous differential inclusions, Proc. AMS 104 (1988), 840-844.

[153] Marano S. A., Classical solutions of partial differential inclusions in Banach spaces, Appl. Anal. 42 (1991), 127-143.

[154] Marano S. A., Generalized solutions of partial differential inclusions depending on a parameter, Rend. Accad. Naz. Sci. XL 13 (1989), 281-295.

[155] McClendon J. F., Subopen multifunctions and selections, Fund. Math. 121 (1984), 25-30.

[156] McLennan A., Fixed points of contractible valued correspondences, Internat. J. Game Theory 18 (1989), 175-184.

[157] Michael E. A., Continuous selections I, Ann. Math. 63 (1956), no. 2, 361-382.

[158] Michael E. A., Continuous selections III, Ann. Math. 65 (1957), no. 2, 374-381.

[159] Michael E. A., A generalization of a theorem on continuous selections, Proc. Amer. Math. Soc. 105 (1989), no. 1, 236-243.

[160] Michael E. A., Continuous selections: A guide for avoiding obstacles, Gen. Topol. and Relat. Mod. Anal. and Algebra, vol. 6, Proc. the Prague Topol. Symp., Aug. 25-29, 1986, Berlin, 1988, 344-349.

[161] Myjak J., A remark on Scorza-Dragoni theorem for differential inclusions, Ćas. Pstováni Mat. 114 (1989), 294-298.

[162] Nasselli Riccieri O., A-fixed points of multivalued contractions, J. Math. Anal. Appl. 135 (1988), 406-418.

[163] Nasselli Riccieri O., Riccieri B., Differential inclusions depending on parametere, Bull. Polish Acad.Sci. Math. 37 (1989), 665-671.

[164] Neustadt L. W., The existence of optimal control in the absence of convexity, J. Math. Anal. Appl. 7 (1963), 110-117.

[165] Nistri P., Obukhovskii, Zecca P., Viability for feedback control systems in Banach spcaes via Carathéodory closed-loop controls, Differential Equations Dynam. Systems 4 (1996), 367-378.

[166] Nguyen Tran cao, A characterization of some weak lower-semicontinuity of integral functionals, Bull. Polish Acad. Sc. Math. 23 (1975).

[167] Nguyen H. T., Semicontinuity and continuous selections for the multivalued superposition operator without assuming growth-type conditions, Studia Math. 163 (2004), 1-19.

[168] Nguyen H. T., Juniewicz M., Ziemińska J., $CM-$Selectors for pairs of oppositely semicontinuous multivalued maps with L_p-decomposable values, Studia Math. 144 (2001), 135-152.

[169] Nowak A., Rom C., Decomposable hulls of multifunctions, Discussiones Math. 22 (2002), 233-242.

[170] Obukhovskii V., On periodic solutions of differential equations with multivalued right-hand side, Trudey Mat. Fak. Voronezh. Gos. Univ., vol. 10, 1973, 99. 74-82 (in Russian).

[171] Olech, C., Lexicographical order, range of integrals and "bang-bang" principle Mathematical Theory of Control, Academic Press New York and London 1967, 35-45.

[172] Olech C., Lectures on the integration of set-valued functions, preprint SISSA, 44 M, 1987.

[173] Olech C., Decomposability as substitute for convexity, in: Multifunctions and integrands, G Salinetti (ed.), Lecture notes in Math. 1091, Springer- Verlag, Berlin, 1984, 193-205.

[174] Olech C., Existence of solutions of non convex orientor field, Boll. Un. Mat. Ital. 4 (1975), 1985-197.

[175] Olech C., Boundary solutions of differential inclusions, Lecture Notes in Math. 979 (1983), 236-239.

[176] Olech C., The characterization of the weak* closure of certain sets of integrable functions, SIAM J. Control 12(2), (1974), 311-318.

[177] Olech C., A necessary and sufficient condition for lower semi-continuity of certain integral functionals, Studia Math.70 (1978).

[178] Olech C., Approximation of set-valued functions by cotinuous functions, Colloq. Math. 19 (1968), 285-293.

[179] Olech C., A note concerning set-valued measurable functions, Bull. Polish Acad. Sci. Math. Astronom. Phys., 13 (1965), 317-321.

[180] Olech C., Extremal solutions of a control system, J. Diff. Equations 2 (1966), 74-101.

[181] Olech C., A note concerning extremal points of a convex set, Bull. Polish Acad. Sci. Math. Astronom. Phys., 13 (1965), 347-351.

[182] Olech C., Integrals of set-valued functions and optimal control problems, Colloque sur la theorie Mathematique du Controle Optimale, Vander, Louvain - Belgique (1970), 109-125.

[183] Olech C., A Contribution to the time optimal control problem, III Konferenz uber nichtlineare Schwingungen, Teil II, Academie-Verlag-Belin, 1966, 438-446.

[184] Olech C., On the range of an unbounded vector-valued measure, Math. System Theory 2 (1968), 251-256.

[185] Olech C., On n-dimensional extensions of Fatou's lemma, J. of Applied Mathematics and Physics (ZAMP), 38 (1987), 266-272.

[186] O'Neil B., A fixed point theorem for multivalued functions, Duke Math. J. 14 (1947), 689-693.

[187] O'Neil B., Essential sets and fixed points, Amer. J. Math. 75 (1953), 497-509.

[188] Ornelas A., A continuous version of the Filippov-Gronwall inequality for differential inclusions, Preprint SISSA, 78M, 1988.

[189] Ornelas A., Nonconvex problems for differential inclusions, Ph. D. Thesis, SISSA 1988.

[190] Oxtoby J., Measure and Category, Springer-Verlag, New York, 1971.

[191] Papageorgiou N. S., A propery of the solution set of differential inclusions in Banach spaces with Carathéodory orientor field, Appl. Anal. 27 (1988), 279-287

[192] Papageorgiou N. S., Boundary value problems for evolution inclusions, Comment. Math. Univ. Carolin. 29 (1988), no. 2, 355-363.

[193] Papageorgiou N. S., Fixed point theorems for multifunctions in metric and vector spaces, nonlinear Anal. 7 (1983), 763-770.

[194] Papageorgiou N. S., On the solution eveolution set of differential inclusions in Banach spaces, Appl. Anal. 25 (1987), 319-329.

[195] Patniak S. N., Fixed points of multiple-valued transformations, Fund. Math. 65 (1969), 345-349.

[196] Plaskacz S., On the solutions sets for differential inclusions, Boll. Un. Mat. Ital. A. 6 (1992), 387-394.

[197] Pliś A., Measurable orientor fields, Bull. Polish Acad. Sci. Math. 13 (1965), 565-569.

[198] Powers M. J., Fixed point theorems for non-compact approximative ANT's, Fund. Math. 75 (1972), 61-68.

[199] Przesławski K., Linear and Lipschitz continuous selectors for the family of convex sets in Euclidean vector spaces, Bull. Polish Acad. Sci. Math. 33 (1985), 31-33.

[200] Przesławski K., Rybiński L., Michael continuous selection theorem under weak lower semicontinuity assumption Proc. Amer. Math. Soc. 109 (1990), 537-543.

[201] Przesławski K., Yost D., Lipschitz retracts, selectors and extensions, Michigan Math. J. 42 (1995), 555-571.

[202] Repovš D., Semenov P. V., Continuous selections of multivalued mappings, Math. Appl. 455, Kluwer, Dordrecht, the Netherlands, 1998.

[203] Ricceri B., Une propriété topologique de l'ensamble des points fixes d'une contraction multivoque à valeurs convexes, Atti Acad. Lincei Rend. Fis. Mat. Natur., LXXXI (1987), 283-286.

[204] Ricceri B., On multifunctions with convex graph, Atti. Accad. Naz. Lincei Rend. Fis. Mat. Natur. 77 (1984), 64-70.

[205] Rockafellar, R.T., Integrals which are convex functionals, Pacific J. Math., 24 (1968), 525-539.

[206] Rockafellar, R.T., Integrals which arę convex functionals. II, Pacific J. Math., 29 (1971), 439-469.

[207] Rockafellar, R.T., Measurable dependance of convex sets and functions on parameters, J. Math. Anal. Appl. 28 (1969), 4-25.

[208] Rybiński L., On Carathéodory type selections, Fund. Math. 125 (1985), 187-193.

[209] Rybiński L., A fixed point approach in the study of solution sets of Lipschitzean functional-differential inclusions, JMAA 160 (1991), 24-46.

[210] Rzeżuchowski T., On the set where all the solutions satisfy a differential inclusion, Qual. Theory Diff. Equations, vol. 2, Amsterdam (1981), 903-913.

[211] Rzeżuchowski T., Scorza-Dragoni type theorem for upper semicontinuous multi-valued functions, Bull. Polish Acad. Sci. Math. 28 (1980), 61-66.

[212] Rzeżuchowski T., Strong convergence results related to strict convexity, Comm. part. Diff. Eqs 9 (1984), 439-466.

[213] Sierpiński W., Sur les fonctions d'ensemble additives and continues, Fund. Math. 3 (1922), 240-246.

[214] Tolstogonov A. A., Differential inclusions in Banach spaces, Soc. Acad. of Sciences, Siberian Branch, Novosibirsk, 1986 (in Russian).

[215] Tolstogonov A. A., On the structure of the solution set for differential inclusions in Banach spaces, Math. USSR Sbornik 46 (1983), 1-15 (in Russian).

[216] Ważewski T., On an optimal control problem, Proc. Conf. Diff. Eq. and their applic. (1964), 229-242.

[217] Valadier M., Closedness in the weak topology of the dual pair L_1, C, J. Math. Anal. Appl. 69 (1979), 17-34.

[218] Yorke J. A., Another proof of Lyapounov convexity theorem, SIAM J. Control p (1971), 351-353.

[219] Yoshida K., Functional analysis, Springer Verlag, Berlin, 1980.

[220] Yost D., There can be no Lipschitz version of Michael's selection theorem, Proc. of the Analysis Conf, Singapore 1986, 295-299.

[221] Zaremba S. K., Sur certaines familles de courbes en relation avec la théorie des équations differentielles, Rocznik Polskiego Tow. Matemat. XV (1963), 83-100.

[222] Zecca P., Stefani G., Multivalued differential equations on manifolds with application to control theory, Illinois J. Math. 24 (1980), 560-575.